土木工程专业英语

Special English for Civil Engineering

王 琼 主编

内容简介

根据土木工程各专业的特点，本着覆盖面广、知识面宽及适当介绍前沿专业知识的原则，本书题材广泛，共包括三个部分：第一部分为土木工程专业英语文献选编，涉及土木工程基本知识，以及建筑工程、工程材料、结构工程、桥梁工程、道路工程、地下工程、岩土工程、工程管理、智能建造等专业领域，内容丰富，材料充实新颖，有效地拓展了专业知识的广度；第二部分为土木工程专业英语写作，能使土木类学生较全面地掌握专业英语写作的基本知识、要求和应用技能，为今后的工作打下良好的基础；第三部分为土木工程专业英语词汇，选编了土木工程专业常用的词汇及术语，增强了本书的适用性。

本书具有知识覆盖面广、应用性强的特点，适用于高等院校土木工程专业本科生的教学，也可以作为从事土木工程领域工作人员的词汇查阅、英语资料阅读及论文书写的参考用书。

图书在版编目(CIP)数据

土木工程专业英语/王琼主编. —哈尔滨：哈尔滨工业大学出版社，2024.5

ISBN 978-7-5767-1400-5

Ⅰ. 土… Ⅱ. ①王… Ⅲ. ①土木工程-英语 Ⅳ. ①TU

中国国家版本馆 CIP 数据核字(2024)第 107529 号

土木工程专业英语

TUMU GONGCHENG ZHUANYE YINGYU

策划编辑 许雅莹
责任编辑 周一曈
封面设计 高永利
出版发行 哈尔滨工业大学出版社
社　　址 哈尔滨市南岗区复华四道街 10 号　邮编 150006
传　　真 0451-86414749
网　　址 http://hitpress.hit.edu.cn
印　　刷 哈尔滨博奇印刷有限公司
开　　本 880 mm×1230 mm　1/32　印张 7.875　字数 231 千字
版　　次 2024 年 5 月第 1 版　2024 年 5 月第 1 次印刷
书　　号 ISBN 978-7-5767-1400-5
定　　价 38.00 元

前　言

随着“一带一路”倡议的深入实施，土木工程领域的国际交流和合作日益频繁。专业英语是拓展本专业知识面的工具，是打开进入本学科知识领域和世界前沿大门的钥匙。根据2023年版《高等学校土木工程本科专业指南》规定：科技与专业外语为土木类学生必须掌握的工具知识。为满足高等学校土木类人才培养的需要，根据土木工程各专业的特点和土木工程专业英语教学的新要求，结合编者多年来的教学实践，编写了本书。

本书以提高土木类学生的专业英语阅读能力为出发点，通过对课文的阅读和学习，巩固专业英语的基础知识，增加专业英语词汇和专业术语，从而具备一定的阅读专业英语文献的能力和翻译技巧，能够以英语为工具，通过阅读去获取与本专业有关的国外科技信息，了解本专业的最新国际动态。

本书具有知识系统、适用广泛等特点。根据土木工程各专业的特点，本着覆盖面广、知识面宽及适当介绍前沿专业知识的原则，本书题材广泛，共包括三个部分：第一部分为土木工程专业英语文献选编，涉及土木工程基本知识，以及建筑工程、工程材料、结构工程、桥梁工程、道路工程、地下工程、岩土工程、工程管理、智能建造等专业领域，内容丰富，材料充实新颖，有效地拓展了专业知识的广度；第二部分为土木工程专业英语写作，能使土木类学生较全面地掌握专业英语写作的基本知识、要求和应用技能，为今后的工作打下良好的基础;第三部分为土木工程专业英语词汇,选编了土木工程专业常用的词汇及术语,增强了本书的适用性。

本书是编者在参考了国内外诸多版本教材的基础上,取长补短,借鉴经验编写而成的。本书的编写参阅了诸多文献资料,在此对这些文献的作者表示衷心感谢。

受编者知识水平所限,疏漏之处在所难免,敬请广大读者不吝赐教。

编　者

2024 年 4 月

CONTENTS

1 Collection of Literature about Civil English

Lesson 1 Civil Engineering

1.1 Introduction of Civil Engineering

Civil engineering is one of the oldest of the engineering specialties. It refers to the technical activities related to engineering materials and equipment involved in engineering construction, such as survey, design, construction and maintenance. It also means the objects of engineering construction, such as houses, roads, railways, transportation pipelines, tunnels, bridges, canals, dams, ports, power plants, airports, offshore platforms, irrigation and drainage system, and protective engineering, which is essential to modern society and large population concentrations.

The word "civil" derives form the Latin for citizen. In 1782, Englishman John Smeaton used the term to differentiate nonmilitary engineering work from that of the military engineering, which was predominant at that time. A civil engineer is one, who designs or constructs public works, such as roads, railways, sewers, bridges, harbors, canals, etc. As opposed to a military engineer, or to a mechanical engineer, who makes machines etc., the field of civil engineering is much broader.

1.2 Scope of Civil Engineering

The scope of civil engineering is so broad that it can be subdivided into many technical specialties. When a project begins, the site is surveyed and mapped by civil engineers. Geotechnical specialists perform soil experiments to determine if the earth can bear the weight of the project. Environmental specialists study the project's impact on the local area: the potential for air and groundwater pollution, the project's impact on local animal and plant life, and how the project can be designed to meet government requirements aimed at protecting the environment. Transportation specialists determine what kind of facilities are needed to ease the burden on local roads and other transportation networks. The structural specialists use preliminary data to make detailed designs, plans and specifications for the project. Supervising and coordinating the work of these civil engineer specialists, from beginning to end of the project, are the construction management specialists. Based on information supplied by the other specialists, construction management civil engineers estimate quantities and costs of materials and labor, schedule all work, order materials and equipment for the job, hire contractors and subcontractors, and perform other supervisory work to ensure the project is completed on time and as specified.

Throughout any given project, civil engineers make extensive use of computers. Computers are used to design the project's various elements and to manage it. Computers are a necessity for the modern civil engineer because they permit the engineer to efficiently handle the large quantities of data needed in determining the best way to construct a project.

1.2.1 Structural engineering

In this speciality, civil engineers plan and design structures of all types, including bridges, dams, power plants, supports for equipment, special structures for offshore projects, the space program, transmission towers, giant astronomical and radio telescopes, and many other kinds of projects. Using computers, structural engineers determine the forces a structure must resist: its own weight, wind and hurricane forces, temperature changes that expand or contract construction materials, and earthquakes. They also determine the combination of appropriate materials: steel, concrete, plastic, stone, asphalt, brick, aluminum, or other construction materials.

1.2.2 Water resources engineering

Civil engineers in this specialty deal with all aspects of the physical control of water. Their projects help prevent floods, supply water for cities and for irrigation, manage and control rivers and water runoff, and maintain beaches and other waterfront facilities. In addition, they design and maintain harbors, canals and locks, build huge hydroelectric dams, smaller dams and water impoundments of all kinds, help design offshore structures, and determine the location of structures affecting navigation.

1.2.3 Geotechnical engineering

Civil engineers who specialize in this field analyze the properties of soils and rocks that support structures and affect structural behavior. They evaluate and work to minimize the potential settlement of buildings and other structures that stems from the pressure of their weight on the earth. These engineers also evaluate and determine how to strengthen the stability of slopes and fills, and how to protect structures against earthquakes and the effects of groundwater.

1.2.4 Environmental engineering

In this branch of engineering, civil engineers design, build and supervise systems to provide safe drinking water and to prevent and control pollution of water supplies, both on the surface and underground. They also design, build and supervise projects to control or eliminate pollution of the land and air. These engineers build water and wastewater treatment plants, and design air scrubbers and other devices to minimize or eliminate air pollution caused by industrial processes, incineration or other smoke-producing activities. They also work to control toxic and hazardous wastes through the construction of special dump sites or the neutralizing of toxic and hazardous substances. In addition, the engineers design and manage sanitary landfills to prevent pollution of surrounding land.

1.2.5 Transportation engineering

Civil engineers working in this specialty build facilities to ensure safe and efficient movement of both people and goods. They specialize in designing and maintaining all types of transportation facilities, highways and streets, mass transit systems, railroads and airfields, ports and harbors. Transportation engineers apply technological knowledge as well as consideration of the economic, political and social factors in designing each project. They work closely with urban planners, since the quality of the community is directly related to the quality of the transportation system.

1.2.6 Pipeline engineering

In this branch of civil engineering, engineers build pipelines and related facilities which transport liquids, gases or solids ranging from coal slurries (mixed coal and water) and semiliquid wastes, to water, oil and various types of highly combustible and noncombustible gases. The engineers determine pipeline design, the economic and environmental impact of a project on regions it must traverse, the type of materials to be used—steel, concrete, plastic or combinations of various materials—installation techniques, methods for testing pipeline strength, and controls for maintaining proper pressure and rate of flow of materials being transported. When hazardous materials are being carried, safety is a major consideration as well.

1.2.7 Construction engineering

Civil engineers in this field oversee the construction of a project from beginning to end. Sometimes called project engineers, they apply both technical and managerial skills, including knowledge of construction methods, planning, organizing, financing and operating construction projects. They coordinate the activities of virtually everyone engaged in the work: the surveyors, workers who lay out and construct the temporary roads and ramps, excavate for the foundation, build the forms and pour the concrete; workers who build the steel framework. These engineers also make regular progress reports to the owners of the structure.

1.2.8 Community and urban planning

Those engaged in this area of civil engineering may plan and develop communities within a city or entire cities. Such planning involves far more than engineering consideration, environmental, social and economic factors in the use and development of land and natural resources are also key elements. These civil engineers coordinate planning of public works along with private development. They evaluate the kinds of facilities needed, including streets and highways, public transportation systems, airports, port facilities, water-supply and wastewater-disposal systems, public buildings, parks and recreational, and other facilities to ensure social and economic as well as environmental well-being.

1.2.9 Photogrametry, surveying and mapping

The civil engineers in this specialty precisely measure the Earth's surface to obtain reliable information for locating and designing engineering projects. This practice often involves high technology methods, such as satellite and aerial surveying, and computer-processing of photographic imagery. Radio signals from satellites, scans by laser and sonic beams, are converted to maps to provide far more accurate measurements for boring tunnels, building highways and dams, plotting flood control and irrigation projects, locating subsurface geologic formations that may affect a construction project, and a host of other building uses.

1.3 Importance of Civil Engineering

Civil engineering, as a very important basic discipline, has its important properties: comprehension, sociability, practicality and unity. Civil engineering provides a corporeal technology basis for the development of the national economy and the improvement of people's life and plays a promoting role in the revitalization of many industries. Engineering construction is the basic production process of fixed assets. Therefore, the construction industry and the real estate have become a pillar of the economy in many countries and regions.

Words and Phrases

civil engineering	土木工程
power plant	发电站
irrigation and drainage system	灌溉及排水系统
geotechnical specialist	岩土专家
construction management	工程管理
contractor and subcontractor	承包商和分包商
structural engineering	结构工程
hurricane	飓风
water resources engineering	水利工程
water runoff	雨水径流
hydroelectric dam	水力发电堤坝
settlement	沉降
stability of slope	边坡稳定性
environmental engineering	环境工程
safe drinking water	安全饮用水
transportation engineering	交通工程
pipeline engineering	管线工程
installation technique	安装技术

rate of flow of material	流体流速
construction method	施工方法
progress report	进度报告
urban planning	城市规划
water-supply and wastewater-disposal system	供水及污水处理系统
aerial survey	航空测量
sonic beam	声束
corporeal technology	实体技术
fixed asset	固定资产

Lesson 2 Materials of Civil Engineering

2.1 Introduction of Building Materials

In searching for economical and efficient solutions to the problems that have been posed, engineers seek effective exploitation of the mechanical properties of the materials that are being used: properties such as stiffness (the change in dimensions as the material is subjected to changes in load) and the strength (the limit to the amount of load that can be supported). Through experience and understanding, engineers have become more adept at exploiting these properties in order to produce increasingly daring structures. Effective exploitation implies that civil engineers are able either to control these properties to suit the application or to understand the nature of the properties of the materials with which they are presented.

For example, the steel and concrete that we see in structures above the ground can be designed to have chosen stiffness and strength. These are manufactured materials which may have to be transported a long way to reach a construction site. The costs of transportation may be regarded as excessive, and locally sourced wood and rock may be available instead—their strength and stiffness are as you find them. Rock is just one constituent of the more general range of materials that make up the ground. The softer materials are designated as soils though the boundary between weak rocks and hard soils is not well defined.

2.2 Natural Construction Materials

2.2.1 Soil

All civil engineering constructions sit upon or sit in the ground. Tunnels pass through the ground; foundations for buildings and bridges are excavated from the ground; aeroplane land on the ground; motor cars and railway trains drive on roads or rails laid on the ground; and ships berth against structures which are attached to the ground. So the behaviour of the soil or other materials that make up the ground in its natural state is rather important to engineers. However, although it can be guessed from exploratory survey and from knowledge of the local geological history, the exact nature of the ground can never be discovered before construction begins. Road embankments are formed of carefully prepared soils, and water-retaining dams may also be constructed from selected soils and rocks, these can be seen as designer soils. Today, tourists flock to Pisa mostly because of the unfortunate incident of the uneven settlement, they come to see the leaning tower (Figure 2.1). The tower is an example of Romanesque construction with semicircular arches round the lower storey and semicircular arcades of arches around each storey of the tower. The few window openings on the staircase which spirals round inside the wall are small. The bells are hung within the top storey.

Figure 2.1 Leaning tower in Pisa

Soils are formed of mineral particles packed together with surrounding voids, and the particles can never pack perfectly. Soils can contain a very wide range of particle sizes depending on the source material. For evaluation, soils are graded according to particle size, which increases from silt to clay to sand to gravel to rock. In general, the larger particle soils will support heavier loads than the smaller ones.

A leaning tower may be a valuable draw for tourists but as the tilt increases, the tower becomes progressively less stable. The daring decision was taken by an international committee of geotechnical engineers (civil engineers with a particular interest in the behaviour of the ground). Modest safety measures were installed in the form of lead weights on the up-tilt side of the foundation together with cables which could have provided some slight restraint if the tilt had started to accelerate. Then small chunks of clay were extracted from under the up-tilt side of the foundation until the tower had rotated backwards to the inclination which it had had in about 1838. Its lifetime was thus successfully extended.

2.2.2 Timber

Timber is one of the earliest construction materials and one of the few natural materials with good tensile properties. Hundreds of different species of wood are found throughout the world, and each species exhibits different physical characteristics. Only a few species are used structurally as framing members in building construction. Softwoods, both because of their abundance and because of the ease with which their wood can be shaped, these species of timber more commonly used in the United States for construction are Douglas fir, Southern pine, spruce and redwood. The ultimate tensile strength of these species varies 5,000 - 8,000 psi (1 psi = 6.895 kPa). Hardwoods are used primarily for cabinetwork and for interior finishes such as floors.

Because of the cellular nature of wood, it is stronger along the grain than across the grain. Wood is particularly strong in tension and compression parallel to the grain, and it has great bending strength. These properties make it ideally suited for columns and beams in structures. Wood is not effectively used as a tensile member in a truss, however, because the tensile strength of a truss member depends upon connections between members. It is difficult to devise connections which do not depend on the shear or tearing strength along the grain, although numerous metal connectors have been produced to utilize the tensile strength of timbers.

However, timber has some disadvantages, such as uneven structure, moisture absorption and water absorption, resulting in large swelling, shrinkage deformation, flammability, and easy to be fire hazardous.

2.2.3 Stone

Stone (or rock) is a good construction material. It is a compressive material, it cannot withstand a tensile force. The ultimate compressive strength of bonded masonry depends on the strength of the masonry unit and the mortar. The ultimate strength will vary 1,000–4,000 psi, depending on the particular combination of masonry unit and mortar used.

Evidently, there are different types of rock with different strengths and different abilities to resist the decay that is encouraged by sunshine, moisture and frost, but rocks are generally strong, dimensionally stable materials: they do not shrink or twist with time. We might measure the strength of a type of rock in terms of the height of a column of that rock that will just cause the lowest layer of the rock to crush. A 150 m high solid pyramid uses quite a small proportion of this available strength. If the rock can be easily cut, then it can be readily used for the carving of shapes and figures.

The builders of the Egyptian pyramids (Figure 2.2) had the task of creating a massive rock structure around a small burial chamber. The small hole created local stress concentrations which they could relieve by leaning slabs of stronger stone against each other to form an internal pitched roof. Or alternatively they could create an internal protection by placing blocks of stone one on top of the other on each side of the void.

The Romans developed the semicircular arch as a structural form. An arch is extremely strong once it is complete, provided it is able to push sideways against unyielding abutments. A little mortar on

Figure 2.2 Egyptian pyramids

the radial joints between the section can provide some extra strength, but is not essential if the stones are carefully prepared. Careful study of masonry structures, old and new, will usually reveal some modest (harmless) cracking from which the mechanism of displacement of the masonry can be deduced.

2.3 Manufactured Construction Materials

2.3.1 Iron and steel

Stone is very strong when compressed or pushed, but not so strong in tension. The provision of iron links between adjacent stone blocks can help to provide some tensile strength. Cast iron can be formed into many different shapes and is resistant to rust, but is brittle, when it breaks it loses all its strength very suddenly. Iron with a low proportion of carbon, is more ductile. It can be stretched without losing all its strength, and can be beaten or rolled (wrought) into simple shapes. Steel is a mixture of iron with a higher proportion of carbon than wrought iron and with other elements (such as manganese or titanium or chromium) which provide particular mechanical benefits. Mild steel has a remarkable ductility—a tolerance of being stretched, which results from its chemical composition and which allows it to be rolled into sheets or extruded into chosen shapes without losing its strength and stiffness. There are limits on the ratio of the quantities of carbon and other elements to that of the iron itself in order to maintain these desirable properties for the mixture.

Steel is an outstanding structural material. It has a high strength on a pound-for-pound basis when compared 10 other materials, even though its volume-for-volume weight is more than ten times that of wood. It has a high elastic modulus, which results in small deformations under load. Steel elements can be joined together by various means, such as bolting, riveting or welding.

2.3.2 Concrete

Concrete is a sort of artificial rock, which can be cast into almost any chosen shape and size. The Romans discovered that mixing pulverize volcanic ash product found in the region with water produced a reaction, which, when complete, left a very strong rock-like material. The dome of the Pantheon in Rome was built in around 126 AD using this early concrete.

Concrete is a mixture of water, sand and gravel, and Portland cement. Crushed stone, manufactured lightweight stone and seashells are often used in lieu of natural gravel. Portland cement, which is a mixture of materials containing calcium and clay, is heated in a kiln and then pulverized. Concrete derives its strength from the fact that pulverized Portland cement, when mixed with water, hardens by a process called hydration. In an ideal mixture, concrete consists of about three fourths sand and gravel (aggregate) by volume and one fourth cement paste. The physical properties of concrete are highly sensitive to variations in the mixture of the components, so a particular combination of these ingredients must be custom-designed to achieve specified results in terms of strength or shrinkage. When concrete is poured into a mold or form, it contains free water, not required for hydration, which evaporates. As the concrete hardens, it releases this excess water over a period of time and shrinks. As a result of this shrinkage, fine cracks often develop. In order to minimize these shrinkage cracks, concrete must be hardened by keeping it moist for at least 5 d. The strength of concrete increases in time because the hydration process continues for years; as a practical matter, the strength at 28 d is considered standard.

Concrete deforms under load in an elastic manner. Although its elastic modulus is one tenth that of steel, similar deformations will result since its strength is also about one tenth that of steel. Concrete is basically a compressive material and has negligible tensile strength. As concrete sets, the chemical reactions that turn a sloppy mixture of cement and water and stones into a rock-like solid, produce a lot of heat. If a large volume of concrete is poured without any special precautions, then as it cools down, having solidified, it will shrink and crack. The Hoover Dam (Figure 2.3) was built as a series of separate concrete columns of limited dimension through which pipes carrying cooling water were passed in order to control the temperature rise. Evidently the junctions between the blocks needed to be sealed and the cooling pipes to be filled, and the rock beneath and beside the dam to be rendered impermeable.

Figure 2.3 Hoover Dam

2.3.3 Reinforce concrete

A French gardener, Joseph Monier, presented his invention—a reinforced concrete pot and a railway sleeper. After that, he continued to search for new materials and obtained patents including reinforced concrete pots and reinforced concrete beams that were applied to highway guardrails. He made it possible to find that reinforcing steel is placed in the tensile zone and the easily crackable area of the concrete member, which can greatly improve the tensile strength and bending strength, and can limit the occurrence and development of cracks. His invention was quickly applied to the construction sector. This process is workable because steel and concrete expand and contract equally when the temperature changes. If this was not the case, the bond between the steel and concrete would be broken by a change in temperature since the two materials would respond differently. Reinforced concrete can be molded into innumerable shapes, such as beams, columns, slabs and arches, and is therefore easily adapted to a particular form of building. In 1875, the world's first reinforced concrete structure, designed by Monier, was built at the Castle of Chazelet, Paris, which is a new era in the history of human architecture.

Reinforced concrete structures have been used extensively in the engineering community since 1900. In 1928, a new type of reinforced concrete structure, prestressed reinforced concrete, appeared and was widely used in engineering practice after the World War Ⅱ. The invention of reinforced concrete and the application of steel in the construction industry in the mid-19th century made it possible to construct tall buildings and long-span bridges.

Words and Phrases

mechanical property	力学性能
stiffness	刚度
strength	强度
load	荷载
adept	擅长于
exploit	开发,利用
nature of the property	性质的本质
manufactured material	人造材料
construction site	施工场地
berth	停泊,泊位
behaviour	性能
natural state	天然状态
exploratory survey	地质勘察
embankment	路堤
Romanesque construction	罗马式建筑
semicircular arch	半圆拱

arcade	拱廊
spiral	旋转
lead weight	铅坠
up-tilt	向上倾斜的
chunk of clay	土块
ultimate tensile strength	极限抗拉强度
column	柱子
beam	梁
shrinkage deformation	收缩变形
masonry unit	砌块
burial chamber	墓室
internal pitched roof	内坡屋顶
radial joint	径向接头
mild steel	低碳钢
Pantheon in Rome	万神殿
Hoover Dam	胡佛水坝
impermeable	不透水的
railway sleeper	铁路轨枕

Lesson 3 Basic Structural Forms of Civil Engineering

The buildings, where people live and work, are composed of individual spaces, each unit space is made up of basic structure components, including deck, beam, column, arch and truss. Materials and structural components are combined to form the various parts of a building, including the load-carrying frame, skin, floors and partitions.

3.1 Deck

Decks refer to a bending component with a larger plane size and a smaller thickness, usually placed horizontally. They are generally used in construction projects for floors, roof, foundation slab, partitions, etc. The deck can be divided into unidirectional board and bidirectional board according to the form of force. When the rectangular plate is supported on both sides, it is a unidirectional plate. When there are four side supports, the load on the board is transmitted to the four sides in both directions, which is called a bidirectional board. When the long side of the board is much longer than the short side, the load on the board is mainly transmitted to the supporting components along the short side direction, while the load transmitted along the long side direction is very small and can be ignored. Such a four sided supporting board is still considered as a unidirectional board.

3.2 Beam

3.2.1 Definition

A beam is a component that bears vertical loads and is primarily subjected to bending. Beams are generally placed horizontally to support decks and withstand various vertical loads transmitted by the decks and their own weight. The beams and decks together form the floor and roof structure of the building. In a frame structure, beams connect with columns in all directions as a whole. In a wall structure, the connecting beam above the void connects two wall limbs to work together. In frame shearing-wall structures, beams play two roles in above two structures. The beam undergoes bending deformation under vertical loads and is subjected to tension on one side, and the other side is under compression. At the same time, shear force is transmitted through mutual displacement between cross-sections, and the vertical load acting on it is ultimately to the supports on both sides. The internal force of a beam includes shear force and bending moment.

3.2.2 Classification of beams

From a functional perspective, there are structural beams, such as foundation beams, frame beams, etc. Together with vertical components such as columns and load-bearing walls, a spatial structural system is formed, consisting of structural beams such as ring beams, lintels, connecting beams, etc., which play a constructive role in crack resistance, earthquake resistance, stability and other aspects. According to the structural engineering properties, beams can be divided into frame beams, frame beams supported by shear walls, inner frame beams, beams, masonry wall beams, masonry lintels, shear wall connecting beams, shear wall concealed beams and shear wall frame beams. According to construction technology, there are cast-in-place beams, prefabricated beams, etc. Beams can be made by different material, such as steel, reinforced concrete, wood. From the state of force, it can be divided into statically determinate beams and statically indeterminate beams. A statically indeterminate beam refers to a beam that remains geometrically invariant and has no additional constraints. The other one refers to a beam that remains geometrically invariant and has additional constraints.

3.3 Column

Columns are one of the most basic components in architecture, playing a role in supporting and distributing loads. According to different architectural styles and uses, columns come in different forms and materials.

3.3.1 Classify by shape

1. Cylinder

A cylinder is the most common form of column, with a circular cross-section. A cylinder can withstand loads applied in any direction and is easier to manufacture.

2. Square column

The cross-section of a square or rectangular column is commonly used in modern architecture. Compared to cylinders, square columns are easier to install and can better adapt to modern architectural design styles.

3. Polygonal column

The cross-section of this kind of column is a polygon, such as a hexagon, octagon, etc. This type of column is mainly used in classical architecture and religious places.

4. Composite column

A composite column is a structure composed of two or more different shapes. For example, in ancient Greece and Rome, composite Greek/Roman columns with a square or rectangular bottom and a cone or sphere at the top were very popular.

3.3.2 Classified by material

1. Stone column

Stone columns are one of the oldest types of columns, made from natural or artificial stone. In ancient architecture, they were the main element used to support buildings.

2. Wooden column

Wooden columns are commonly used in residential and lightweight buildings. Wood can be decorated through carving and painting.

3. Steel column

Steel columns are commonly used in high-rise buildings and large-span structures. Steel has extremely high strength and stiffness, and can withstand a large number of loads.

4. Concrete column

Concrete columns are one of the most common types of columns in modern architecture. Concrete has excellent compressive performance and is easy to process into various shapes.

5. Glass fiber reinforced plastic (FRP) column

FRP materials have high strength and durability, and are relatively lightweight. In recent years, FRP materials have been widely used in the construction field.

3.3.3 Classified by function

1. Structural support column

Structural support columns are the main elements used to support buildings. This type of column is commonly used in large buildings and bridges.

2. Decorative column

Decorative columns are mainly used to beautify the appearance of buildings. This type of pillar is usually decorated through carving, painting and other methods.

3. Functional column

Functional columns are a design that perfectly combines functionality and aesthetics. For example, in shopping malls or public places, columns can be designed to have both supporting and decoration functions.

3.4 Arch

3.4.1 Characteristics of arch

An arch is a curved structure characterized by horizontal support reactions under vertical loads.

The characteristics of arches is as followings.

(1) The bending moment is smaller than the corresponding simply supported beam, and the reason for the existence of horizontal

thrust.

(2) Low material consumption, light weight and large span.

(3) Brick and stone materials with strong compressive performance can be used.

(4) The structure is complex and the construction cost is high.

3.4.2 Application of arch

According to different classification methods, arches can be divided into three hinge arches: double hinge arches, non hinge arches and tie rod arches. Arch structures are widely used in arch bridge construction, including circular lintels for brick masonry doors and windows in brick concrete structures, as well as large arched structures. Zhaozhou Bridge (Figure 3.1) is the world's oldest and most intact large-span single span open shoulder flat arch bridge.

Figure 3.1 Zhaozhou Bridge

The world's largest span dual-purpose steel truss arch bridge, for both public and railway, is Changtai Changjiang River Bridge (Figure 3.2).

Figure 3.2 Changtai Changjiang River Bridge

3.5 Truss

A truss is a geometric shape invariant structure composed of triangular frames made up of straight rods. The joint point between members is called a node. According to the axis of the truss members and the distribution of external forces, trusses can be divided into planar trusses and spatial trusses. A spatial structure such as a roof truss or bridge is composed of a series of parallel planar trusses. If they mainly bear plane loads, they can be simplified as plane trusses for calculation. The truss components are generally made of steel and connected by welding, riveting and bolting.

The Tianxingzhou Changjiang Special Channel Bridge (Figure 3.3) has a total length of 727 m and a main span of 388 m, spanning the 269-meter-wide Tianxingzhou Special Channel. It adopts the structural form of a steel truss arch bridge. 40,000 t of steel were used in the construction, which is close to the total steel consumption of the Beijing Olympic Stadium Bird's Nest. The entire bridge uses over 1,200 rods of different lengths and types, with up to 26 closure points forming 14 closure points. Each point needs to achieve millimeter level docking, with high accuracy requirements.

Figure 3.3 Tianxingzhou Changjiang Special Channel Bridge

Words and Phrases

truss	桁架
load-carrying frame	承载框架
skin	表层
partition	隔墙
plane	平面
unidirectional board	单向板
vertical load	竖向荷载
limb	边缘
shearing-wall	剪力墙
mutual	相互的,交叉的
ring beam	环梁,圈梁
lintel	过梁
connecting beam	结合梁

statically indeterminate	超静定的
invariant	不变的
constraint	约束
cross-section	横截面
polygonal column	多角柱
hexagon	六边形
octagon	八边形
cone	圆锥
sphere	球形
aesthetics	美学
horizontal thrust	水平推力
light weight	轻质
tie rod arch	拉杆拱
spatial truss	空间桁架

Lesson 4 Architecture

Architecture is a spatial environment constructed by people using natural or artificial materials, following certain scientific principles and artistic rules. The so-called artificial refers to the construction activities that require materials, technology and workers during the construction process of buildings. Architecture not only provides people with a concealed internal space, but also creates a different external space in the field of civil engineering. There are two terms need to be differentiated: building and constructed works. The building refers to a space where people can work, live and study, such as common residences, schools, shopping malls, etc. However, constructed works do not have such space, such as embankments, water towers, etc.

4.1 The Three Elements of Architecture

4.1.1 Building function

Building function refers to the fact that a building should meet people's usage requirements. For example, it should have a comfortable environment, appropriate space and reasonable layout, advanced, high-quality and convenient facilities, etc., so that people can live comfortably, work smoothly, and rest happily inside.

4.1.2 Technology and economic conditions

Construction is a product of technical engineering. Like mechanical and hydraulic engineering, it is achieved for a certain purpose through material and engineering technology. Therefore, architecture is a material product that requires a significant amount of material and energy consumption during construction and use, as well as the provision of relevant technology and equipment conditions. Buildings are limited by these technological and economic conditions.

4.1.3 Building image

A building satisfies not only people's material needs, but also their spirit. Architecture gives people a spiritual feeling through its color, texture and spatial form.

4.2 The Form of Building

The form of building is an outgrowth of its function, its environment and various socioeconomic factors.

4.2.1 The influence of function

An apartment building, an office building and a school differ in form because of the difference in the functions they fulfill. In an apartment building, all habitable spaces, such as living rooms and bedrooms, must have natural light from windows while bathrooms and kitchens can have artificial light and therefore car be in the interior of the building. This set of requirements places a natural limit on the depth of an apartment building. In office buildings, on the other hand, artificial light is accepted for more uniform illumination and therefore the depth of such buildings is not limited by a need for natural light.

4.2.2 The influence of environment

Environment may affect both the shape and appearance of a building. An urban school may create its own environment by using blank walls to seal out the city completely, and a country school may develop as an integral part of the landscape, even though both schools fulfill the same function.

4.2.3 The influence of socioeconomic factor

Finally, the form of a building is affected by a variety of socioeconomic factors, including land costs, tenancy, building budget and building restrictions. High land costs in urban areas result in high buildings, while low land costs in the country result in low buildings. A housing project for the rich will take a more different form than a low-cost housing project. A prestige building will be more generously budgeted for than other office buildings. The bulk of a building and its outline may be limited by building restrictions. In all these examples, buildings with similar functions take on different forms.

4.3 Aesthetic Principles of Architectural Form

4.3.1 Unification and change

Buildings objectively have factors of unity and change. For example, rooms with the same functional use in a building are treated uniformly in terms of floor height, bay, doors, windows, etc. Some buildings also adopt uniform geometric shapes. The different components, materials and appearance reflect changing factors.

4.3.2 Equilibrium and stability

Equilibrium means that parts of a building, such as front, rear, sides, should have sufficient corresponding to each other. The human eye is accustomed to a balanced combination, which easily gives people a sense of stability, equilibrium and completeness. On common building facades, a symmetrical architectural form on both sides is a typical example of balance.

But symmetry is not the only way to achieve equilibrium. There can also be asymmetric and dynamic equilibrium (Figure 4.1, Figure 4.2). The layout of Shanghai Jinmao Building is mainly octagonal, with local variations. The basic form of the Sydney Opera House is similar, but there are also certain differences, with both unified continuity and certain changes in form.

Figure 4.1 Shanghai Jinmao Building

Stability is focus on the overall relationship between the top and bottom of a building, such as light to heavy and small to big.

Figure 4.2 Sydney Opera House

However, with the emergence of new materials and structures, some shapes that break through the traditional concept of stability have also been favored by people.

4.3.3 Proportion and scale

Architecture calls for good proportions—a pleasing relationship of voids to solids, of height to width, of length to breadth. The coordinated proportion can give people a beautiful feeling. In architectural design, efforts are made to achieve a symmetrical height and appropriate width. Many attempts have been made to explain good proportions by mathematical formulas, such as the golden section.

The scale involves specific dimensions and accurately expresses the size relationship between the building and the human body, as well as the size relationship between various parts of the building. It should visually convey its true size.

4.4 Conclusion

Through the related architectural forms shaped by their purpose, governed by the materials. Proportioned and given scale and character by the designer, buildings become expressions of the ideals and aspirations of the generations that built them. The successive styles of historic architecture are incarnations of the spirit of their times.

Words and Phrases

conceal	隐藏
embankment	筑堤;堤,路堤,坝体,填方
texture	质地
spatial form	空间形态
socioeconomic factor	社会经济因素
habitable space	居住空间
depth	进深
illumination	照明
land cost	土地价格
building budget	建筑预算
building restriction	建筑限制条例
architectural form	建筑形式
unification	统一
floor height	层高
bay	开间,跨度
geometric shape	几何形状
equilibrium	平衡
facade	立面
symmetry	对称
Sydney Opera House	悉尼歌剧院
mathematical formula	数学公式

golden section	黄金分割
aspiration	野心
historic architecture	历史建筑
incarnation	体现

Lesson 5 Building Engineering

The purpose of a building is to provide a shelter for the performance of human activities. From the time of the cave dwellers to the present, one of the first needs of men has been a shelter from the elements. The purpose of building engineering is to meet people's various needs such as living, working and entertainment. A comfortable, safe and fully functional building environment can be created by building engineering, which also improve the development level of cities and communities, promotes economic prosperity and social progress.

5.1 Components of a Building

A building is composed of such elements as foundations, walls or columns, floors, roofs, stairs, doors, windows, balconies, and so on (Figure 5.1). These components can be divided into two categories: building enclosure (skin) and load-bearing structures.

5.1.1 Skin

The function of a building is achieved through an enclosure

Figure 5.1 Components of a building

structure, which forms a closed area to prevent external wind, rain and snow from invading. However, it allows sunlight and fresh air to enter, allowing people to live and work safely and comfortably. Walls, doors, windows, eaves, fences and canopies can be seen as skin. Windows and doors are the two most important types of enclosure structures in architecture. Under different usage conditions, it also has functions, for example thermal or sound insulation, waterproofing, fire prevention. They are often made of materials such as aluminum alloy, plastic, glass, steel, etc.

5.1.2 Load-carrying frame

A building is subjected to weight of itself, occupant and equipment, structural materials, as well as to the effects of snow, wind, earthquake, all of which are loads. The structure in a building that can withstand these loads and transmit them to other parts, is called a load-carrying structure. It is composed of various elements such as beams, slabs, columns, roof trusses, load-bearing walls, foundations, etc. The load acting on the building is from top to bottom during design.

1. Foundation

The foundation is the lowest part of a building, it will bear the loads of the entire building and transmit it to the soil.

2. Wall

According to location, walls can be divided into exterior and interior walls. The exterior wall is the external enclosure structure that protects the rooms from elements. The function of the interior wall is to divide the space on each floor into multiple areas or rooms. Based on different structural performance, there are load-bearing and non load-bearing walls. The former can withstand loads from roofs or floors, while other walls are called non load-bearing walls. Partitions, infill walls and curtain walls are the non load-bearing examples.

3. Floor

The construction of the floors in a building depends on the basic structural frame that is used. In steel skeleton construction, floors are either slabs of concrete resting on steel beams or a deck consisting of corrugated steel with a concrete topping. In concrete construction, the floors are either slabs of concrete on concrete beams or a series of closely spaced concrete beams (ribs) in two directions topped with a thin concrete slab, giving the appearance of a waffle on its underside.

4. Roof

Roof is the top part of a building, it is not only an enclosure structure that prevents the space beneath it from being invaded by wind, rain and snow, but also a load-carrying structure that bears the weight of the roof and various live loads. Flat, slope and curve are the three basic shapes of a roof, which must be constructed with a certain slope to discharge rainwater. A flat roof is with a slope of 10% or less.

5.2 Classification by Height

5.2.1 Multi-story building

Multi-story and high-rise structures are mainly used in residential shopping malls, office buildings, hotels and other buildings. The boundary of above two varies from country to country. In China, buildings with less than 8 floors are called multi-story buildings, while those with 8 floors or more are called high-rise buildings. The commonly used structural forms for multi-layer structures are combination structures and frame structures.

5.2.2 Tall building

High-rise buildings are a product of modern economic development and technological progress. The world's first modern high-rise building was the Home Insurance in Chicago, USA, with 55-meter height, 10 storeys, and was built in 1883. The development of high-rise buildings has gone through more than 100 years of history, and economics and industrial technology progress have continuously created favorable conditions for it. The high-rise buildings in the city is an important symbol to economic and social prosperity. Since the 1990s, tall buildings have made rapid progress globally.

5.3 Classification by Load Carrying Components

Structural systems can be divided into frame structure, shear wall structure, frame-shear structure and tube structure.

5.3.1 **Frame structure**

The load-carrying system of a frame structure is composed of beams and columns, which is reasonable for bearing vertical loads. However, its ability to withstand horizontal loads is poor, so it is only suitable for use when the height or storey number of the building is not large. Because, in this case, the influence of wind loads is minimal, and vertical loads play a controlling role in the design of the structure. The reasonable number of layers for a framework is 6 – 15, with the most economical being around 10 layers. The general ratio of height to width is about 5–7.

5.3.2 **Shear wall structure**

Reinforced concrete walls, used to withstand vertical and horizontal loads, are called shear wall structures. They also serve as a enclosure and partition. Due to the direct transmission of vertical loads from floors or roofs to shear walls, the spacing between shear walls depends on the span of the floor slab. In general, the spacing between shear walls is 3 – 8 m, which is suitable for residential buildings and hotels.

5.3.3 **Frame-shear structure**

When a portion of shear walls are installed in a frame structure, it is named frame-shear structure(Figure 5.2). The frame and shear walls work together to carry loads, jointly bearing vertical and horizontal loads. The frame and shear wall work together, the latter bears the majority of horizontal loads, while the former bears vertical loads, which can greatly reduce the cross-section of the column. This structure system belongs to a semi-rigid structural system.

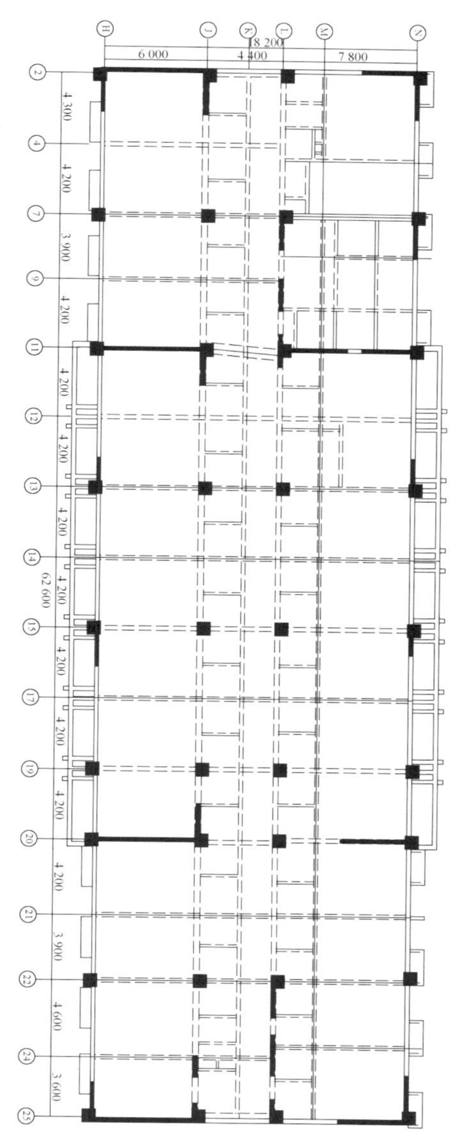

Figure 5.2 Frame-shear structure(unit: mm)

5.3.4 Tube structure

A tube structure is a high-rise building system consisting of one or more tubes as load-bearing structures. Under the action of lateral wind loads, the force acting on it is similar to that of a rigid box shaped cantilever beam. The windward side will be under tension, while the leeward side will be compressed.

1. Frame tube structure

This system is composed of outer frame tubes and internal general frames. The frame tube can be made of steel or reinforced concrete material. The frame tube serves not only as the spatial anti lateral force system of the building, but also as the enclosure wall of the building, directly forming windows between beams and columns. This structural form was first proposed by the famous American engineer Fazler R. Khan. In 1963, the first building to adopt a framed tube structure was built in Chicago: the 43-storey Dwight Chestnut Apartments. The Shenzhen Hualian Building adopts a frame tube structure system (Figure 5.3). The building was built in 1997, with a height of 88 m and a total of 22 floors.

2. Tube in tube system

When the layout of the surrounding frame columns is dense, the surrounding frame can be regarded as the outer tube, while the shear wall of the inner core can be regarded as the inner tube, forming a tube in a tube system.

3. Bundled tube

With the continuing need for larger and taller buildings, the tubes may be used in a bundled form to create larger tube envelopes while maintaining high efficiency. The 110-storey Sears Roebuck Headquarters Building (Figure 5.4) in Chicago has nine tubes, bundled at the base of

Figure 5.3 Shenzhen Hualian Building(unit: m)

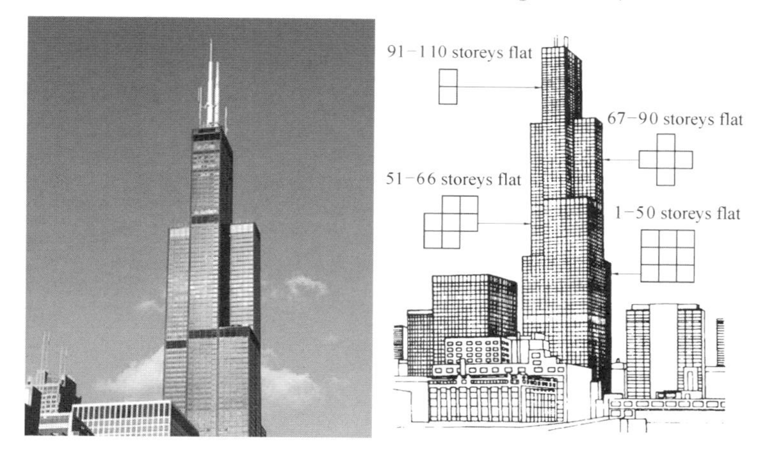

Figure 5.4 Sears Roebuck Headquarters Building

the building in three rows. Some of these individual tubes terminate at different heights of the building, demonstrating the unlimited architectural possibilities of this latest structural concept. The Sears Roebuck Headquarters Building, at a height of 1,450 ft (1 ft=30.48 cm), is the world's tallest building at that time.

Words and Phrases

the time of the cave dwellers	穴居人时代
elements	元素、天气
economic prosperity	经济繁荣
social progress	社会发展
building enclosure	建筑物围护结构
invade	侵入
eave	屋檐,茅草檐
canopy	天篷(棚),雨棚,棚
sound insulation	隔音
waterproofing	防水
fire prevention	防火
aluminum alloy	铝合金
load-bearing wall	承重墙
infill wall	填充墙
curtain wall	幕墙
rib	肋
flat roof	平屋顶
combination structure	混合结构,组合结构
frame structure	框架结构
tube structure	筒体结构
residential building	居民楼
frame-shear structure	框剪结构
semi-rigid	半刚性的
tube in tube system	筒中筒体系
bundled tube	成束管

Lesson 6 Highway Engineering

6.1 Importance of Highway Engineering

Highways have played a key role in the development and sustainability of human civilization from ancient times to the resent. Today, throughout the world, highways continue to dominate the transportation system—providing critical access for the acquisition of natural resources, industrial production, retail marketing and population mobility. The influence of highway transportation on the economic, social and political fabric of nations is far-reaching and, as a consequence, highways have been studied for decades as a cultural, political and economic phenomenon. While industrial needs and economic forces have clearly played an important part in shaping highway networks, societies' fundamental desire for access to activities and affordable land has generated significant highway demand, which has helped define and shape highway networks.

In the 21 century, the role of highways in the transportation system continues to evolve. In most nations, the enormous investment in highway-transportation infrastructure that occurred in the middle of the last century, has now given way to infrastructure maintenance and rehabilitation, improvements in operational efficiency, various traffic-congestion relief measures, energy conservation, improved safety and environmental mitigation. This shift has forced a new emphasis in highway engineering and traffic analysis—one that requires a new skill set and a deeper understanding of the impact of highway decisions.

6.2 Highways Planning

6.2.1 Basic requirements

As for highways, the basic requirements for the planning and layout of the network are as follows.

(1) Highway network planning must work closely with other transportation networks to form a coordinated comprehensive transportation. Since the layout of highway routes is less objectively restricted and more flexible than railways and water transportation, it is necessary to create convenient conditions for the connection and development of railways and water transportation as much as possible.

(2) The technical grade of the trunk highway network, traffic facilities along the lines and their construction sequences should be planned according to the importance of the passing area and the size of the traffic volume.

(3) It should make full use of the original roads and local roads, and gradually improve to meet the road network level and technical standard requirements through improvement.

(4) It should conform to the principles of phased construction and engineering economy.

(5) It should strive to meet the requirements of low road network density, short transportation lines, high transportation efficiency and

low transportation costs.

(6) Highway network planning should also pay attention to the needs of local farmland water conservancy construction and development of local resources.

(7) The highway grade should be determined comprehensively according to the planning of the highway network and the long-term traffic volume, starting from the overall situation and combining the tasks and nature of the highway.

(8) The environmental protection is required. Whether in the process of road construction, environmental protection, or vehicle exhaust gas, noise, and road sewage discharge and diversion during operation, it should be fully considered in the planning process.

6.2.2 The form of the highway network

The form of the highway network generally depends on the following factors.

(1) The transportation needs between administrative and economic centers.

(2) The size and direction of passenger and cargo transportation flows.

(3) The natural conditions of the planned area, especially the distribution of mountains, the direction of large rivers and negative engineering geological conditions.

(4) Special requirements for national defense, etc.

6.3 Classification of Highway

According to the *Technical Standard of Highway Engineering by the Ministry of Transport* of the People's Republic of China (abbreviated as *Technical Standard*), the highway is divided into five technical grades: motorway, 1st-class highway, 2nd-class highway, 3rd-class highway and 4th-class highway. There are still some rural highways below the 4th-class.

(1) The motorway is a multi-lane highway exclusively for motor vehicles to drive in different directions and lanes, and is access controlled. The design annual average daily traffic volume (AADT) of motorways should be more than 15,000 passenger cars.

(2) The 1st-class highway is a multi-lane highway where motor vehicles can drive in different directions and lanes, and access can be controlled as needed. The design annual average daily traffic volume of 1st-class highways should be more than 15,000 passenger cars. That is to say, the 1st-class highway has an equivalent design traffic volume to motorway.

(3) The 2nd-class highway is a two-lane road for motor vehicles. The design annual average daily traffic volume of 2nd-class highways should be 5,000–15,000 passenger cars.

(4) The 3rd-class highway is a two-lane highway for mixed driving of motor vehicles and non-motor vehicles. The annual average daily traffic volume of 3rd-class highways should be 2,000–6,000 passenger cars.

(5) The 4th-class highways are two-lane or single-lane highways for mixed driving of motor vehicles and non-motor vehicles. The annual average daily traffic volume of two-lane 4th-class highways should be less than 2,000 passenger cars; the annual average daily traffic volume of single-lane 4th-class highways should be less than 400 passenger cars.

6.4 Base Course Material

The base course of the pavement plays a role of linking the

surface and the road base in the pavement structure. According to the difference in stiffness, the material is divided into three types: flexible base, semi-rigid base and rigid base. According to the type of binder, it is divided into non-binding gravel materials, inorganic binding semi-rigid materials, and asphalt (organic) binding material of asphalt stabilized crushed stone.

6.4.1 Unbonded granular base

Crushed stone refers to the stone material that meets the requirements of the project, which is mined and processed according to a certain size. Gravel refers to the non-angular granular material that is transported by water for a long time. Graded crushed stone refers to a material composed of crushed stone designed according to a certain grading requirement. Graded gravel refers to a material composed of gravel designed according to certain gradation.

6.4.2 Inorganic binder stabilized base

The physical and mechanical properties of inorganic binder stabilized materials include stress-strain relationship, fatigue characteristics and shrinkage (temperature shrinkage and dry shrinkage) characteristics. Therefore, inorganic binder stabilized materials have the characteristics of good stability, strong frost resistance and self-contained slabs of the structure itself, but their wear resistance is poor, so they are widely used in the construction of pavement structure base and subbase. The pavement base course built by this is the base course of inorganic binder stabilized material, or called semi-rigid base course, including lime stabilized base, cement stabilized base and industrial waste residue stabilized base.

6.4.3 Asphaltic base

Asphalt binder mixture refers to a mixture composed of asphalt, coarse, fine aggregates and mineral powder, and designed according to a certain mix ratio design method. It is mixed, paved, rolled and formed into a pavement structure as a base course, which is called an asphalt binder base course. According to different design air void and usage, asphalt binder mixtures can be divided into asphalt treated base (ATB, design air void of 6% – 39%, used as base course), semi-open asphalt stabilized macadam (asphalt macadam, AM, with a design air void of 6% – 12%, used as a low-grade road surface) and open-graded asphalt stabilized gravel (used for pavement drainage of design air void of 18% – 22%, including asphalt treated permeable base, referred to as ATPB, used for base drainage). The mix design and application process of the asphalt binder base is basically the same as that of asphalt concrete, and physical and mechanical properties of the material are also very similar.

6.4.4 Cement concrete base

Lean concrete is a kind of concrete made by mixing coarse and fine aggregates with a certain amount of cement and water. This kind of concrete has a lower cement content than ordinary concrete. Compared with commonly used semi-rigid materials such as cement stabilized crushed stone, it has higher strength, rigidity and integrity, and is resistant to erosion and frost. It has good fatigue performance, yields to rigid base materials, and is close to cement concrete pavement in nature. The material composition design and construction mainly refer to cement concrete.

6.5 Highways and the Human Element

It is important to keep in mind that highway transportation is part of a larger transportation system that includes air, rail, water and pipeline transportation. In this system, highways are the dominant mode of most passenger and freight movements. For passenger travel, highways account for about 90% of all passenger-miles. On the freight side, commercial trucks account for about 37% of the freight ton-miles and, because commercial trucks transport higher-valued goods than other modes of transportation (with the exception of air transportation), nearly 80% of the dollar value of all goods is transported by commercial trucks. So highways play a dominant role in both passenger and freight movement. It is critical to plan, design, construct and maintain the highways.

Words and Phrases

acquisition	收获,获得
natural resource	自然资源
industrial production	工业产品
retail marketing	零售市场,零售营销
fabric	织物,结构
infrastructure	基础设施
rehabilitation	修复工程
traffic-congestion	交通拥堵
mitigation	缓解
flexible	灵活的
original road	原有公路
phased construction	分期建设

water conservancy	水利
traffic volume	交通量
road sewage discharge	道路污水排放
diversion	临时支路
administrative and economic center	行政和经济中心
negative engineering geological condition	不良工程地质条件
technical standard	技术标准
motorway	高速公路
rural highway	农村公路
base course	基层
pavement	路面
road base	路基
gradation	级配
frost resistance	抗冻性
subbase	底基层
binder	黏合剂
waste residue	残渣
lean concrete	贫混凝土,少灰混凝土

Lesson 7 Bridge Engineering

7.1 History of Bridges

A bridge is a structure built to span physical obstacles without closing the way underneath such as a body of water, valley or road, for the purpose of providing passage over the obstacles, usually something that can be detrimental to cross otherwise. There are many different designs that each serve a particular purpose and apply to different situations. Designs of bridges vary depending on the function of the bridge, the nature of the terrain where the bridge is constructed and anchored, the material used to make it, and the funds available to build it.

The very first bridges were built by Mother Nature and were the result of falling trees spanning streams or canyons. Man used these accidental crossings to reach new areas for settlement and new sources of food. The first man-made bridges copied this and were made from logs and then later stones. These very early designs relied on the log or stone being long enough to cross the gap in a single span. Later designs used simple cross beams and support to cross larger gaps.

There are six main types of bridges: beam bridges, cantilever bridges, arch bridges, suspension bridges, cable-stayed bridges and truss bridges.

7.2 Beam Bridges

Beam bridges are horizontal beams supported at each end by piers (Figure 7.1). The earliest beam bridges were simple logs across streams and similar simple structures. In modern times, beam bridges are large box steel girder bridges. Weight on top of the beam pushes straight down on the piers at either end of the bridge. They are made up mostly of wood or metal.

Figure 7.1 Beam bridge

The beam bridge, also known as a girder bridge, is a firm structure that is the simplest of all the bridge shapes. Both strong and economical, it is a solid structure composed of a horizontal beam, being supported at each end by piers that endure the weight of the bridge and the vehicular traffic. Compressive and tensile forces act on a beam bridge, due to which a strong beam is essential to resist bending and twisting because of the heavy loads on the bridge. When traffic moves on a beam bridge, the load applied on the beam is transferred to the piers. The top portion of the bridge, being under compression, is shortened; while the bottom portion, being under tension, is consequently stretched and lengthened. Trusses made of steel are used to support a beam, enabling dissipation of the compressive and tensile forces. In spite of the reinforcement by trusses, length is a limitation of a beam bridge due to the heavy bridge and truss weight. The span of a beam bridge is controlled by the beam size since the additional material used in tall beams can assist in the dissipation of tension and compression.

Extensive research is being conducted by several private enterprises and the state agencies to improve the construction techniques and materials used for the beam bridges. The beam bridge design is oriented towards the achievement of light, strong and long-lasting materials like reformulated concrete with high performance characteristics, fiber reinforced composite materials, electro-chemical corrosion protection systems and more precise study of materials. Modern beam bridges use prestressed concrete beams that combine the high tensile strength of steel and the superior compression properties of concrete, thus creating a strong and durable beam bridge. Box girders are being used that are better designed to undertake twisting forces, and can make the spans longer, which is otherwise a limitation of beam bridges. The modern technique of the finite element analysis is used to obtain a better beam bridge design, with a meticulous analysis of the stress distribution, and the twisting and bending forces that may cause failure.

7.3 Cantilever Bridges

A cantilever bridge is a bridge built using cantilevers structures that project horizontally into space, supported on only one end (Figure 7.2). For small footbridges, the cantilevers may be simple beams; however, large cantilever bridges designed to handle road or rail traffic use trusses built from structural steel, or box girders built from prestressed concrete. The steel truss cantilever bridge was a major engineering breakthrough when first put into practice, as it can span distances of over 460 m, and can be more easily constructed at difficult crossings by virtue of using little or no falsework.

Figure 7.2 Cantilever bridge

A simple cantilever span is formed by two cantilever arms extending from opposite sides of the obstacle to be crossed, meeting at the center. In a common variant, the suspended span, the cantilever arms do not meet in the center; instead, they support a central truss bridge which rests on the ends of the cantilever arms. The suspended span may be built off-site and lifted into place, or constructed in place using special traveling supports.

A common way to construct steel truss and prestressed concrete cantilever spans is to counterbalance each cantilever arm with another cantilever arm projecting the opposite direction, forming a balanced cantilever. Thus, in a bridge built on two foundation piers, there are four cantilever arms: two arms which span the obstacle, and two anchor arms which extend away from the obstacle. Because of the need for more strength at the balanced cantilever's supports, the bridge superstructure often takes the form of towers above the foundation piers. The Commodore Barry Bridge is an example of this type of cantilever bridge.

Steel truss cantilevers support loads by tension of the upper members and compression of the lower ones. Commonly, the structure distributes the tension via the anchor arms to the outermost supports, while the compression is carried to the foundations beneath the central towers. Many truss cantilever bridges use pinned joints and are therefore statically determinate with no members carrying mixed loads.

Prestressed concrete balanced cantilever bridges are often built using segmental construction. Some steel arch bridges are built using pure cantilever spans from each side, with neither falsework below nor temporary supporting towers and cables above. These are then joined with a pin, usually after forcing the union point apart, and when jacks are removed and the bridge decking is added the bridge becomes a truss arch bridge. Such unsupported construction is only possible where appropriate rock is available to support the tension in the upper chord of the span during construction, usually limiting this method to the spanning of narrow canyons.

7.4 Arch Bridges

An arch bridge is a bridge with abutments at each end shaped as a curved arch. Arch bridges work by transferring the weight of the bridge and its loads partially into a horizontal thrust restrained by the abutments at either side. A viaduct may be made from a series of arches, although other more economical structures are typically used today. There are some variations of arch bridges.

7.4.1 Corbel arch bridges

The corbel arch bridge (Figure 7.3) is a masonry or stone bridge where each successively higher course cantilevers slightly more than the previous course. The steps of the masonry may be trimmed to make the arch have a rounded shape. The corbel arch does not produce thrust, or outward pressure at the bottom of the arch, and is not considered a true arch. It is more stable than a true arch because it does not have this thrust. The disadvantage is that this type of arch is not suitable for large spans.

7.4.2 Aqueducts and canal viaducts

In some locations, it is necessary to span a wide gap at a relatively high elevation, such as when a canal or water supply must span a valley. Rather than building extremely large arches, or very tall supporting columns, a series of arched structures are built one atop another, with wider structures at the base (Figure 7.4). Roman civil engineers developed the design and constructed highly refined structures using only simple materials, equipments, and mathematics. This type is still used in canal viaducts and roadways as it has a pleasing shape, particularly when spanning water, as the reflections of the arches form a visual impression of circles or ellipses.

Figure 7.3 Corbel arch bridge

Figure 7.4 Aqueduct and canal viaduct

7.4.3 Deck arch bridges

This type of bridge comprises an arch where the deck is completely above the arch (Figure 7.5). The area between the arch and the deck is known as the spandrel. If the spandrel is solid, usually the case in a masonry or stone arch bridge, it is call a closed-spandrel arch bridge. If the deck is supported by a number of vertical columns rising from the arch, it is known as an open-spandrel arch bridge. The Alexander Hamilton Bridge is an example of an open-spandrel arch bridge.

Figure 7.5 Deck arch bridge

7.4.4 Through arch bridges

This type of bridge comprises an arch which supports the deck by means of suspension cables or tie bars (Figure 7.6). The Sydney Harbour Bridge is a through arch bridge which uses a truss type arch. These through arch bridges are in contrast to suspension bridges which use the catenary in tension to which the aforementioned cables or tie bars are attached and suspended.

Figure 7.6 Through arch bridge

7.4.5 Tied arch bridges

Also known as a bowstring arch, tied arch bridge incorporates a tie between two opposite ends of the arch. The tie is capable of withstanding the horizontal thrust forces which would normally be exerted on the abutments of an arch bridge(Figure 7.7).

Figure 7.7 Tied arch bridge

7.5 Suspension Bridges

A suspension bridge is a type of bridge in which the derrick is hung below suspension cables on vertical suspenders (Figure 7.8). This type of bridge dates from the early 19th century, while bridges without vertical suspenders have a long history in many mountainous parts of the world.

Figure 7.8 Suspension bridge

This type of bridge has cables suspended between towers, plus vertical suspender cables that carry the weight of the deck below, upon which traffic crosses. This arrangement allows the deck to be level or to arc upward for additional clearance. Like other suspension bridge types, this type often is constructed without falsework.

The suspension cables must be anchored at each end of the bridge, since any load applied to the bridge is transformed into a tension in these main cables. The main cables continue beyond the pillars to deck-level supports, and further continue to connections with anchors in the ground. The roadway is supported by vertical suspender cables or rods, called hangers. In some circumstances, the towers may sit on a bluff or canyon edge where the road may proceed directly to the main span, otherwise the bridge will usually have two smaller spans, running between either pair of pillars and the highway, which may be supported by suspender cables or may use a truss bridge to make this connection. In the latter case, there will be very little arc in the outboard main cables.

The main forces in a suspension bridge of any type are tension in the cables and compression in the pillars. Since almost all the force on the pillars is vertically downwards and they are also stabilized by the main cables, the pillars can be made quite slender, as on the Severn Bridge, near Bristol, England. In a suspended deck bridge, cables suspended via towers hold up the road deck. The weight is transferred by the cables to the towers, which in turn transfer the weight to the ground.

Assuming a negligible weight as compared to the weight of the deck and vehicles being supported, the main cables of a suspension bridge will form a parabola. One can see the shape from the constant increase of the gradient of the cable with linear distance, this increase in gradient at each connection with the deck providing a net upward support force. Combined with the relatively simple constraints placed upon the actual deck, this makes the suspension bridge much simpler to design and analyze than a cable-stayed bridge, where the deck is in compression.

7.6 Cable-Stayed Bridges

A cable-stayed bridge is a bridge that consists of one or more columns, with cables supporting the bridge deck (Figure 7.9). There are two major classes of cable-stayed bridges: in a harp design, the cables are made nearly parallel by attaching cables to various points on the tower so that the height of attachment of each cable on the tower is similar to the distance from the tower along the roadway to its lower attachment; in a fan design, the cables all connect to or pass over the top of the towers.

Figure 7.9 Cable-stayed bridge

Compared with other bridge types, the cable-stayed bridge is optimal for spans longer than typically seen in cantilever bridges and shorter than those typically requiring a suspension bridge. This is the range in which cantilever spans would rapidly grow heavier if they were lengthened, and in which suspension cabling does not get more economical were the span to be shortened.

7.7 Truss Bridges

A truss bridge is a bridge composed of connected elements which may be stressed from tension, compression, or sometimes both in response to dynamic loads(Figure 7.10). Truss bridges are one of the oldest types of modern bridges. A truss bridge is economical to constructowing to its efficient use of materials.

Figure 7.10 Truss bridge

Truss girders, lattice girders or open web girders are efficient and economical structural systems, since the members experience essentially axial forces and hence the material is fully utilized. Members of the truss girder bridges can be classified as chord members and web members. Generally, the chord members resist overall bending moment in the form of direct tension and compression and web members carry the shear force in the form of direct tension or compression. Due to their efficiency, truss bridges are built over wide range of spans. Truss bridges compete against plate girders for shorter spans, against box girders for medium spans and cable-stayed bridges for long spans.

For short and medium spans, it is economical to use parallel chord trusses such as Warren truss, Pratt truss, Howe truss, etc., to minimize fabrication and erection costs. Especially for shorter spans, the Warren truss is more economical as it requires less material than either the Pratt or Howe trusses. However, for longer spans, a greater depth is required at the centre and variable depth trusses are adopted for economy. In case of truss bridges that are continuous over many supports, the depth of the truss is usually larger at the supports and smaller at mid-span.

7.8 Comprehension of Bridge

The principle portions of a bridge may be said to be the substructure and the superstructure. This division is used here simply for convenience. Since in many bridges, there is no clear diving line between the two.

Bridges are great symbols of mankind's conquest of space. The sight of the crimson tracery of the golden gate bridge against a setting sun in the Pacific Ocean, or the arch of the Garabit Viaduct soaring triumphantly above the deep gorge, fills one's heart with wonder and admiration for the art of their builders. They are the enduring expressions of mankind's determination to remove all barriers in its pursuit of a better and freer world.

Words and Phrases

physical obstacles 自然障碍
terrain 地带
funds 资金
Mother Nature 大自然
canyons 峡谷
pier 墩
girder bridge 板梁桥
prestressed concrete 预应力混凝土
finite element analysis 有限元分析

meticulous analysis	细致的分析
stress distribution	应力分析
footbridge	人行桥
falsework	脚手架
cantilever arm	(桥或梁的)悬臂部分,悬臂距
counterbalance	均衡重,平衡重,平衡力;抵消
pinned joint	铰节点
segmental construction	分段施工
upper chord	上弦
abutment	支座,桥台
horizontal thrust	水平推力
viaduct	高架桥
trim	砌筑
aqueduct and canal viaduct	水道运河拱桥
spandrel	拱肩
through arch bridge	中承式拱桥,贯通式拱桥
catenary	悬链
tied arch bridge	下承式拱桥
bowstring arch	弓弦式拱桥
suspension bridge	悬索桥
hanger	吊桥
parabola	抛物线
gradient	坡度
cable-stayed bridge	斜拉桥,斜张桥
truss bridge	桁架桥
lattice girder	格构梁

web girder	网格梁
conquest	征服
freer	更加自由的

Lesson 8 Geotechnical Engineering

Geotechnical engineering is a branch of civil engineering concerned with the engineering behavior earth materials. Geotechnical engineering is important in civil engineering, but is also used by military, mining, petroleum or any other engineering concerned with construction on or in the ground. Geotechnical engineering uses principles of soil mechanics and rock mechanics to investigate subsurface conditions and materials; determine the relevant physical/mechanical and chemical properties of these materials; evaluate stability of natural slopes and man-made soil deposits; assess risks posed by site conditions; design earthworks and structure foundations; and monitor site conditions, earthwork and foundation construction.

8.1 Geotechnical Engineering History

Humans have historically used soil as a material for flood control, irrigation purposes, burial sites building foundations, and as construction material for buildings. First activities were linked to irrigation and flood control, as demonstrated by traces of dams, and canals dating back to at least 2000 BC that were found in ancient

Egypt, ancient Mesopotamia and the Fertile Crescent. As the cities expanded, structures were erectly supported by formalized foundations; ancient Greeks notably constructed pad footings and strip-and-raft foundations. Until the 18th century, however, no theoretical basis for soil design had been developed and the discipline was more of an art than a science, relying on past experience.

In the 19th century, Henry Darcy developed what is now known as Darcy's Law, describing the flow of fluids in porous media. Joseph Boussinesg (a mathematician and physicist) developed theories of stress distribution in elastic solids that proved useful for estimating stresses along depth in the ground. William Rankine, an engineer and physicist, developed an alternative to Coulomb's earth pressure theory. Aber Atterberg developed the clay consistency indices that are still used today for soil classification. Oshorne Reynolds recognized in 1885 that shearing causes volumetric dilation of dense and contraction of loose granular materials.

Modern geotechnical engineering is said to have begun in 1925 with the publication of *Erdbaumechanik* by Karl Terzaghi (a civil engineer and geologist). Considered by many to be the father of modern soil mechanics and geotechnical engineering, Terzaghi developed the principle of effective stress, and demonstrated that the shear strength of soil is controlled by effective stress. Terzaghi also developed the framework for theories of bearing capacity of foundations, and the theory for prediction of the rate of settlement of clay layers due to consolidation. Donald Taylor recognized that interlocking and dilation of densely packed particles contributed to the peak strength of a soil in his 1948 book. The interrelationship between volume change behavior (dilation, contraction and consolidation) and shearing behavior were all connected via the theory of plasticity.

Critical state soil mechanics is the basis for many contemporary advance constitutive models describing the behavior of soil.

8.2 Soil Mechanics

In geotechnical engineering, soils are considered a three-phase material composed of rocks (or mineral particles), water and air. The voids of a soil, the spaces in between mineral particles, contain the water and air.

The engineering properties of soils are affected by four main factors: the predominant size of the mineral particles, the type of mineral particles, the grain size distribution, and the relative quantities of mineral, water and air present in the soil matrix. Fine particles (fines) are defined as particles less than 0.075 mm in diameter.

Some of the important properties of soils that are used by geotechnical engineers to analyze site conditions and design earthworks, retaining structures, and foundations are as followings.

8.2.1 Unit weight

Total unit weight: cumulative weight of the solid particles, water and air in the material per unit volume. Note that the air phase is often assumed to be weightless.

8.2.2 Porosity

Ratio of the volume of voids (containing air, water or other fluids) in a soil to the total volume of the soil. A porosity of 0 implies that there are no voids in the soil.

8.2.3 Void ratio

It's the ratio of the volume of voids to the volume of solid particles in a soil. Void ratio is mathematically related to the porosity.

8.2.4 Permeability

A measure of the ability of wafer to flow through the soil, expressed in units of velocity.

8.2.5 Compressibility

The rate of change of volume with effective stress. If the pores are filled with water, then the water must be squeezed out of the pores to allow volumelric compression of the soil. This process is called consolidation.

8.2.6 Shear strength

The shear stress that will cause shear failure.

8.2.7 Atterberg limits

Atterberg limits is concluded of liquid limit, plastic limit and shrinkage limit. These indices are used for estimation of other engineering properties and for soil classification.

8.3 General Considerations for Classification of Soils

It has been stated that soil can be described as gravel, sand, silt and clay according to grain size. Most of the natural soils consist of a mixture of organic material in the partly or fully decomposed state. The proportions of the constituents in a mixture vary considerably and there is no generally recognized definition concerning the percentage of, for instance, clay particles that a soil must have to be classified as clay, etc.

When a soil consists of the various constituents in different proportions, the mixture is then given the name of the constituents that appear to have significant influence on its behavior, and then other constituents are indicated by adjectives. Thus a sandy clay has most of the properties of a clay but contains a significant amount of sand.

The individual constituents of a soil mixture can be separated and identified as gravel, sand, silt and clay on the basis of mechanical analysis. The clay mineral that is present in a clay soil is sometimes a matter of engineering importance. According to the mineral present, the clay soil can be classified as kaolinite, montmorillonite or illite. The minerals present in a clay can be identified by either X-ray diffraction or differential thermal analysis.

Buildings, bridges, dams, etc. are built on natural soils (undisturbed soils), whereas earthen dams for reservoirs, embankments for roads and railway lines, foundation bases for pavements of roads and airports are made out of remolded soils. Sites for structures on natural soils for embankments, etc., will have to be chosen first on the basis of preliminary examinations of the soil that can be carried out in the field. An engineer should therefore be conversant with the field tests that would identify the various constituents of a soil mixture.

The behavior of a soil mass under load depends upon many factors such as the properties of the various constituents present in the mass, the density, the degree of saturation, the environmental conditions, etc. If soils are grouped on the basis of certain definite principles and rated according to their performance, the properties of a given soil can be understood to a certain extent, on the basis of some simple tests.

8.4 Field Identification of Soils

The methods of field identification of soils can conveniently be discussed under the headings of coarse-grained and fine-grained soil materials.

8.4.1 Coarse-grained soil materials

The coarse-grained soil materials are mineral fragments that may be identified primarily on the basis of grain size.

The different constituents of coarse-grained materials are sand and gravel. The size of sand varies 0.075–4.75 mm and that of gravel 4.75–80 mm. Sand can further be classified as coarse, medium and fine. The engineer should have an idea of the relative sizes of the grains in order to identify the various fractions. The description of sand and gravel should include an estimate of the quantity of material in the different size ranges as well as a statement of the shape and mineralogical composition of the grains. The mineral grains can be rounded, subrounded, angular or subangular. The presence of mica or a weak material such as shale affects the durability or compressibility of the deposit. A small magnifying glass can be used to identify the small fragments of shale or mica. The properties of a coarse grained material mass depend also on the uniformity of the sizes of the grains. A well-graded sand is more stable for a foundation base as compared to a uniform or poorly graded material.

8.4.2 Fine-grained soil materials

Inorganic soils are the constituent parts of fine-grained materials, and the silt and clay fractions. Since both these materials are microscopic in size, physical properties other than grain size must be used as criteria for field identification. The classification tests used in the field for preliminary identification are dry strength test, shaking table test and plasticity test.

8.5 Classification of Soils

Soils in nature rarely exist separately as gravel, sand, silt, clay or organic matter, but are usually found as mixtures with varying proportions of these components. Grouping of soils on the basis of certain definite principles world help the engineer to rate the performance of a given soil either as a sub-base material for roads and airfield pavements, foundations of structures, etc. The classification or grouping of soils is mainly based on one or two index properties of soil which are described in detail in earlier sections. The methods that are used for classifying soils are based on one or the other of the following two broad systems.

(1) A textural system which is based only on grain size distribution.

(2) The systems that are based on grain size distribution and limits of soil.

Many systems are in use that are based on grain size distribution and limits of soil.

Words and Phrases

petroleum 石油

soil mechanics 土力学

rock mechanics 岩石力学

natural slope 边坡

earthwork 土方工程

flood control	控制洪水
Mesopotamia	美索不达米亚
strip-and-raft foundation	条形基础
Darcy's Law	达西定律
Coulomb's earth pressure theory	库伦土压力理论
soil classification	土体分类
volumetric dilation	体积膨胀
Erdbaumechanik	地球力学
effective stress	有效应力
consolidation	固结
critical state	临界状态
three-phase material	三相材料
unit weight	容重
porosity	孔隙率
void ratio	孔隙比
permeability	渗透
compressibility	压缩性
Atterberg limit	(黏性土稠度的)阿太堡界限
liquid limit	液限
plastic limit	塑限
shrinkage limit	缩限
grain size	粒径
kaolinite	高岭石
montmorillonite	蒙脱石
illite	伊利石
X-ray diffraction	X 射线衍射

undisturbed soils	未扰动土
microscopic	微小的
dry strength test	干土强度实验
shaking table test	振动台试验
plasticity test	塑性实验

Lesson 9 Underground Engineering

The populations grow in dense urban city centers, so does the demand for space and natural resources. To solve this problem, it has been to build denser and taller buildings in addition to transporting an ever increasing abundance of resources into the city, such as raw materials, water, energy and food, while moving waste back out. This has major implications for livable cities, which in future policy terms might be considered to include aspects of wellbeing, resource security, carbon reduction.

Urban underground space encompasses structures with various functions: storage(e. g. food, water, oil, industrial goods, waste); industry (e. g. power plants); transport (e. g. railways, roads, pedestrain tunnels); utilities and communications (e. g. water, sewerage, gas, electric cables); public use(e. g. shopping centers, hospitals, civil defense structures); private and personal use(e. g. car garages).

9.1 Plan of Underground Space

Effective planning for underground utilization should be an

essential precursor to the development of major underground facilities. This planning must consider long-term needs while providing a framework for reforming urban areas into desirable and effective environments in which to live and work. If underground developments to provide the most valuable long-term benefits possible, then effective planning of this resource must be conducted. Unfortunately, it is already too late for the near-surface zones beneath public rights-of-way in older cities around the world. The tangled web of utilities commonly found is due to a lack of coordination and the historical evolution in utility provision and transit system development.

The underground has several characteristics that make good planning especially problematical.

(1) Difficult to dismantle.

Once underground excavations are made, the ground is permanently altered. Underground structures are not as easily dismantled as surface buildings.

(2) Stability of the excavation.

An underground excavation may effectively reserve a larger zone of ground required for the stability of the excavation.

(3) Investigation data.

The underground geologic structure greatly affects the types, sizes and costs of facilities that can be constructed, but the knowledge of a region's subsurface can only be inferred from a limited number of site investigation and previous records.

(4) Massive investments.

Large underground projects may require massive investments with relatively high risks of construction problems, delays and cost overruns.

In Tokyo, for example, the first subway line (Ginza Line) was installed as a shallow line (10 m deep) immediately beneath the existing layer of surface utilities. As more subway lines have been added, uncluttered zones can only be found at the deeper underground levels. The new Keikyo line in Tokyo is 40 m deep. A new underground super highway from Marunouchi to Shinjuku has been proposed at a 50-meter depth. For comparison, the deepest installations in London are at approximately a 70-meter depth although the main complex of works and sewers is at less than 25 m. Compounding these issues of increasing demand is the fact that newer transportation services (such as the Japanese Shinkansen bullet trains or the French TGV) often require larger cross-section tunnels, straighter alignments and flatter grades. If space is not reserved for this type of use, very inefficient layouts of the underground beneath urban areas can occur.

9.2 Underground Elements in Design

9.2.1 Time

Unlike a surface structure, or more than in any other field of civil engineering, success depends on practical experience. The design of underground structures is necessarily based on simple empirical rules, but these rules can be used safely only by the engineer who has a background of experience. Large projects involving unusual features may call for extensive application of scientific methods to design, but the program for the required investigations cannot be laid out wisely, nor can the results be interpreted intelligently, unless the engineer in charge of design possesses a large amount of experience. Underground structures possess not only the usual three dimensions of length, width and height, but also the role of fourth dimension—time. It is pertinent and important.

9.2.2 The interaction of the host medium

Most of all, the interaction of the host medium with the underground structure plays a prominent role in the proper functioning of an underground structure. Underground structure is not an independent structure acted upon by well-defined loads, and its deformation is not governed by its own internal elastic resistance. The loads acting on an underground structure are ill defined, and its behavior is governed by the properties of the surrounding ground. Design of an underground structure is not a structural problem, but a ground-structure interaction problem, with the emphasis on the ground. For example, after a tunnel is made, to accommodate the tunnel, the host medium undergoes a period of adjustment. And during construction, ground conditions at the tunnel heading involve both transverse arching and longitudinal arching or cantilevering from the unexcavated face. All ground properties are time-dependent, particularly in the short time. Another essential factor for underground structure is the undesirable groundwater flow. With respect to an underground structure, with time the low-pH water seeping through some host ground created large cavities resulting in the collapse of an underground structure which required heavy resource investment to bring the underground structure back into operation.

9.2.3 Combined actions

The interaction between the host medium and the underground structure, the difficulties in assessing the constitutive relationship of the host medium, the ground characteristics change due to undesirable ground-water flow, the techniques depend on homogeneity for extrapolation of ground parameters, the time-space effect after an excavation is made. The restricted space and environment for conducting construction activities, make the final product, the underground structure, costly and time consuming. The design, construction and instrumentation of an underground structure, therefore, require prudent planning, design and construction sequencing and much more complicated than that required for surface structures.

The design of an underground structure, different from that of the surface structures, may not fully depend on calculation because there are a lot of factors which influence on the mechanical properties of the host ground, due to its movements of geological structure during a long period of time, and these factors have not been recognized yet. So the results from theoretical calculation are quite distinct from those of the real situation and are hard to be adopted as a reference of design in practice. Under present conditions, the design of an underground structure do and will still, to a great extent, relies upon experiences and in situ measurements, and it will be, therefore, a task of top priority to develop and improve the models for the design of an underground structure.

9.3 Design Load

For underground structures, the most important loading comes from the host ground itself. In competent host ground, the ground loading on the underground structure is quite insignificant and maybe equal to zero whereas in incompetent ground, it may be quite significant. The host ground pressures on the underground structure is quite complex. It is dependent on several factors, such as the relative stiffness of the structure and the host ground, the elapsed time between the excavation and installation of support, the characteristics of the host ground, the insitu pressures, the size of the opening, the location of water table, and the adopted methods of construction.

9.3.1 Type of load

The loading of an underground structure is rather arbitrarily divided by the civil engineer into three categories, including dead loading, dynamic loading and live loading.

1. Dead loading

Dead loading denotes a constant load on an underground structure due to the weight of the supported structure itself, such as the deadweight of a structure, pressure of rock or soil, and groundwater pressure, etc.

2. Dynamic loading

The short-term load placed on an underground structure during installation, and it changes in the direction or degree of force during operation, such as the pressure produced by explosion wave or earthquake wave.

3. Live loading

A moving, variable weight added to the dead load or intrinsic weight of an underground structure or vehicle, such as the floor load (people or equipment), crane load, rockfall load and other temporary loads.

In addition, design loads of an underground structure also include the internal forces resulting from concrete shrinkage, temperature fluctuation and different settlement.

9.3.2 Lateral earth pressure

The lateral earth pressure refers to the earth pressure against lateral supports such as retaining walls or the bracing in open cuts, with the resistance of the earth against lateral displacement, with the bearing capacity of footings, and with the stability of slopes. Proper design and construction of these structures require a thorough knowledge of the lateral forces that act between the underground structures and the soil masses being retained, and these lateral forces are caused by lateral earth pressure.

1. Earth pressure at rest

Earth pressure at rest is shown in Figure 9. 1. The mass is bounded by a frictionless wall that extends to an infinite depth. A soil element located at a depth z is subjected to effective vertical and horizontal pressures respectively. For this case, since the soil is dry, we have

$$\sigma_c = \gamma z \tag{9.1}$$

where σ_c is effective vertical pressure; γ is unit weight of soil. Also, note that there are no shear stresses on the vertical and horizontal planes.

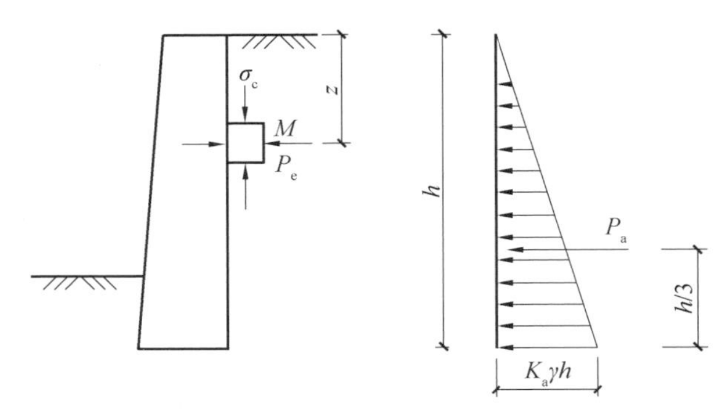

Figure 9. 1 Earth pressure at rest

If the wall is static—that is, if it does not move either to the right or to the left of its initial position—the soil mass will be in a state of elastic equilibrium. That is, the horizontal strain is zero. The ratio of the effective horizontal stress to the vertical stress is called the coefficient of earth pressure at rest, k_0, we have

$$p_0 = k_0 \sigma_c = k_0 \gamma z \tag{9.2}$$

where p_0 is effective horizontal pressure; k_0 is an empirical coefficient, its value depends on the relative density of the soil, the process by which the deposit was formed, and its subsequent stress history.

2. Rankine's earth pressure theory

Rankine investigated the stress conditions in soil at a state of plastic equilibrium. Figure 9. 2 shows the Rankine's active state and Rankine's passive state. It is bounded by a friction-less wall that extends to an infinite depth. The stress condition in the soil element can be represented by the Mohr's circle Ⅱ in Figure 9. 2. However, if the wall is allowed to move away from the soil mass gradually, then the horizontal effective principal stress will decrease. Ultimately a state will be reached at which the stress condition in the soil element can be represented by the Mohr's circle Ⅰ, the state of plastic equilibrium, and failure of the soil will occur. This state is Rankine's active state, and the pressure on the vertical plane is Rankine's active earth pressure.

Following is the expression for Rankine's active pressure. If vertical effective over burden pressure, we have

$$p_a = \gamma z \tan^2\left(45 - \frac{\varphi}{2}\right) - 2c\tan\left(45° - \frac{\varphi}{2}\right) \tag{9.3}$$

Rankine's passive state is also illustrated in Figure 9. 2. The reference wall is a frictionless wall that extends to an infinite depth. The initial stress condition on a soil element is represented by the Mohr's circle Ⅲ.

At this time, failure of the soil will occur. This is referred to as Rankine's passive state, the effective lateral earth pressure, which is

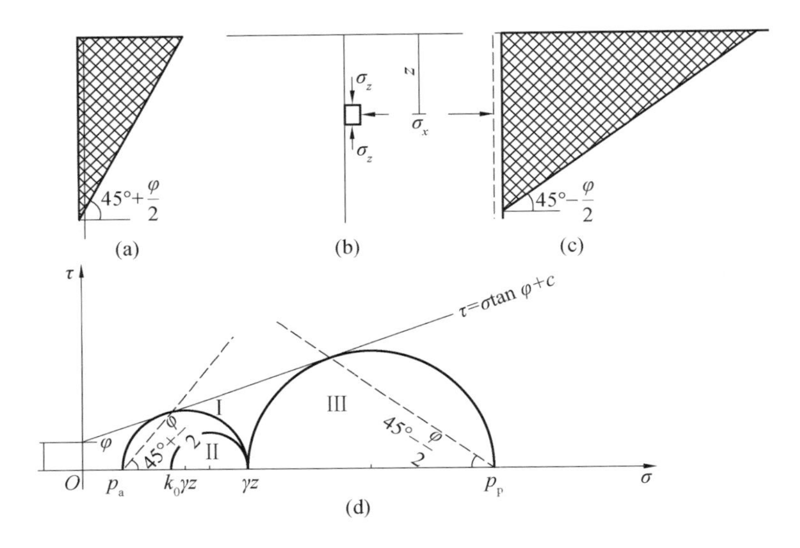

Figure 9.2 Rankine's active state and Rankine's passive state

the major principal stress, is called Rankine's passive earth pressure. From Figure 9.2, it can be shown that

$$p_p=\gamma z\tan^2\left(45°+\frac{\varphi}{2}\right)+2c\tan\left(45°+\frac{\varphi}{2}\right) \tag{9.4}$$

For cohesionless soils, $c=0$, we have

$$p_p=\gamma z\tan^2\left(45°+\frac{\varphi}{2}\right) \tag{9.5}$$

9.4 The Future of Urban Underground Engineering

Underground space utilization can help solve the environmental/resource dilemma in several ways. Underground facilities are typically energy conserving in their own right. More importantly, by using underground space, higher urban densities can be supported with less impact on the local environment. In addition to the obvious benefit of preserving green space and agricultural land, there is strong evidence that higher urban density can lower fuel resource consumption.

Although existing underground facilities throughout the world provide some models for future development, they are all limited in scale, in use, or in their lack of a comprehensive vision for the total city environment. As a complement to more detailed planning and research studies, it is useful to examine the visions of extensive underground complexes, even entire cities that have been proposed by futuristic planners and designers.

Words and Phrases

raw material	原材料
storage	仓储
pedestrain tunnel	行人通道
electric cable	电缆
civil defense structure	人防结构
site investigation	现场勘察
investment	投资
overrun	超支
uncluttered zone	整洁地带
flatter grade	平坦的坡度
background of experience	工程背景
pertinent	直接相关的
interaction	相互作用
undergo	经历,承受
adjustment	调整
unexcavated face	未开挖面
time-dependent	时变性

seep	渗透
homogeneity	同质性
prudent	谨慎的
geological structure	地质结构
host ground	上部地层
elapsed time	经历时间
stability of slope	边坡稳定性
earth pressure at rest	静止土压力
elastic equilibrium	静定平衡状态
active state	主动状态
passive state	被动状态
principal stress	主应力
vision	视野

Lesson 10 Subway Engineering

Subway engineering is a branch of transportation relating to planning, general layout, detailed design construction and operation of subways. These underground systems are major elements in mass transportation. As metropolitan areas increase in size and population, and vehicular traffic become more congested, subways are being given increased consideration in planning urban mass transit systems.

10.1 History

The first subway opened in London in 1863. Steam locomotives fueled by coke and coal were used to pull the trolley cars. In 1896, the first subway on the European continent was placed in service. In Boston, Massachusetts subway lines were instituted in 1895 and 1897. Since then, subways have been constructed in several United States cities.

The most modern subway trains are designed for high-speed travel with maximum safety in comfortable air-conditioned cars. Train movements are controlled by automatic equipment. Since construction of a subway involves putting a railway of special design underground, most of the engineering specialists relating to railways are required in building a subway.

However, in subway work, there is an emphasis on extensive and costly tunnel construction, with tunnels often being at sufficient depth to pass under bodies of water; underground passenger terminals; elaborate ventilation facilities; lengthy underground electric power distribution systems; lighting facilities, escalators for transporting passengers to and from street level; noise control; and safe and reliable signaling facilities.

10.2 Planning

Engineering planning as related to subways includes comprehensive studies to determine whether a subway system is economically feasible. This work involves extensive analysis to evaluate forecasts of construction costs, passenger volumes, passenger fares and revenue, operating costs, depreciation and maintenance of equipment, and passenger safety.

10.3 Construction

10.3.1 Construction methods

Often difficult to install, subways are constructed by either the open-cut method, the tunneling method, or both.

With the open-cut method, a deep open trench is excavated. This frequently requires shoring and bracing at both sides of the trench, in addition to heavy cross framing at different levels. Open-trench procedures limit the depth at which work can be performed.

Tunneling permits subway to be installed at great depths. However, the operation can be costly due to the type of soil or foundation material encountered unstable conditions, and extensive flows of water.

10.3.2 Difficulties during construction

The work of building subways is further complicated by existing underground water supply mains, storm drainage facilities, sanitary sewers, and conduits for electric services and telephone lines. Frequent vehicular traffic along the street under which the subway must be located may add further problems. Often it is difficult to detour traffic on a busy roadway and to close the route to traffic for a long period during subway construction. In such cases a modified open-trench procedure may be followed. Heavy steel vertical pilings are driven along the sides of the area where work must be done. Strong horizontal timbers or sheeting are inserted behind the pilings, and heavy transverse steel beams are installed between the pilings at frequent intervals. Strong timbers or concrete members are then placed longitudinally between the steel beams to serve as a temporary roadway to permit resumption of street traffic.

10.4 Underpinning

10.4.1 Nearby structure

Another factor that makes subway construction difficult is nearby buildings and other structures, especially if these are of sufficient weight to cause heavy loads on the underlying foundation material and large pressures in the area of the proposed subway. Frequently, heavy underpinning must be installed to support adjacent buildings during subway construction. This underpinning may consist of steel pilings and heavy horizontal members on which buildings can be supported.

Large concrete columns reinforced with steel are also used as supports. Columns that are as large as 4–5 ft in diameter have been used. During the installation of any underpinning and the following subway construction activity, it is generally necessary to reduce vibration and settlement of the foundation materials to a minimum.

The design of the bay area rapid transit (BART) system included a difficult but satisfactory method to provide subway service under the harbor between San Francisco and Oakland. Dry-dock sections about 300 ft long made of large-diameter concrete tubes were constructed. These sections were built with temporary closures at the ends so that they could be floated out into the bay and lowered with large rigs into a previously prepared trench in the bay bottom. Each section was joined to the last one placed. Once the sections were in position, the temporary closures at the ends were removed. The procedures in building the BART system were so successful that they were adopted for constructing a 1 mi (1 mi = 1.6 km) section of subway under the HongKong harbor.

10.4.2 Above structures

Determination of the best location for a subway requires much careful study and consideration of various general plans, including underpinning. In some metropolitan area, subway is built underground in the central business area and above ground at other locations. The subway in the southern part of Chicago was placed along an interstate highway at ground level in the median strip, the space dividing opposing traffic on the highway. On another Chicago line leading westward from the central business district, the subway was constructed in a tunnel below the strip of the Congress Street Expressway.

10.5 New Systems

Because of "downtown" congestion and the difficulty of providing adequate parking, one major subway system is the 98 mi network serving Washington, D. C. Extensive auto parking facilities near the several terminals are an integral part of the project. In Atlanta, the 54 mi rail transportation system includes many miles of bus routes.

New subway systems or major additions to older facilities are planned or have been placed in service in several other American cities, including Baltimore, Houston, Los Angeles, Philadelphia and Pittsburgh. Numerous countries have completed or have under construction subway systems. Many foreign countries are served by urban subway networks.

Words and Phrases

steam locomotive 蒸汽机车

coke 焦炭

coal 煤炭

trolley car 无轨电车

ventilation facility 通风设施

escalator 自动扶梯

signaling facility 信号设施

passenger volume 客运量

passenger fare and revenue 乘客票价和收入

open-cut method 明挖法

storm drainage facility 雨水排放设施

sanitary sewer 下水道

detour traffic 绕行交通

heavy steel vertical piling	重型钢立柱
proposed subway	拟建地铁
heavy underpinning	托换
dry-dock section	干坞
rig	钻井设备
interstate highway	州际公路

Lesson 11 Intelligent Construction

11.1 Construction Management Software

The possible use of computer applications in construction lies on following several aspects.

11.1.1 Estimating

Computers can help estimations in most phases of the work. In developing crews and evaluating the productivity of labor and equipment, computers can be the first to provide data from files of project and second to assist with specialized engineering programs for calculations such as cable way cycles, earth-moving engineering, simulation, calculations, etc. In the costing phase, a computer can maintain files of current cost information, and with remote access, it is now becoming possible to tie into various databases maintained by traditional estimating forms. As the estimate nears completion, a computer can be particularly valuable for developing spreadsheets and in bid-sheet preparation.

11.1.2 Cost control

In larger construction companies, computers were often first installed under the control of accounting department. Therefore, it was not surprising that cost control systems were generally extensions of the payroll, and accounting and finance people. Some of the more sophisticated systems apply the principles of engineering economy to include the time value of money in project decision-making.

11.1.3 Scheduling

Critical path scheduling was one of the earliest applications of computers in construction, but has taken time for it to gain wide acceptance. Like cost control, this is an application that should be directly in the hands of management, so the slow acceptance may be a result of the fact that earlier computers were not easily used by non-data processing people.

11.1.4 Quality assurance

Quality assurance applications can begin with the online retrieval of specifications, codes and standards. Quality assurance systems also assist in the documentation of procedures and testing requirements and in the reporting of test results and completion of administrative steps to various interested agencies and parties. Some of the most advanced applications involved not only administrative procedures, but also direct production control.

11.1.5 Purchase

Simpler procurement systems are just extensions of the accounting department accounts payable program. However, in large organizations, there are separate programs for procurement scheduling and expediting to be sure that step such as requisitions and shop drawings are not overlooked or to bring problems and delays to management attention before they become too acute. Materials and procurement systems can also include simple or sophisticated inventory control systems for job materials, tools and supplies.

11.1.6 Executive administration

Even the simpler personal microcomputers can be almost immediately paid for themselves in various kinds of administrative applications. For example, numerous lists must be maintained on projects for drawing logs, tool inventories, safety equipment, etc. Any number of microcomputer file systems can handle such applications. Word processing software on small or large computers can assist with letters and the documentation.

11.1.7 Productivity

Deliberate detail and systematic efforts to improve productivity on construction projects are becoming increasingly common. Computers can assist with the statistical analysis of the questionnaires distributed to workers and supervisors, with the simulation of operations before they are implemented, etc. This is a new area, but with the improved availability of computer on job sites, it is expected to become common in the future.

11.2 Computer-Aided Drafting and Design

11.2.1 Introduction

CADD is an acronym for computer-aided drafting and design.

Initially, the letters CAD referred to computer-aided (or assisted) drafting, but now they can mean computer-aided drafting, computer-aided design, or both. CADD (with two Ds) better describes current technology since many popular software packages include both two-dimensional drafting and three-dimensional design functions. In addition, many CADD systems provide analysis capabilities, schedule production and reporting. All are part of the design process.

11.2.2 Function of CADD

CADD has dramatically changed the way that designers develop and record their ideas. Even though the principles of graphic communication have not changed the method of creating, manipulating, and recording design, concepts have changed with computers. The techniques of drafting, organization of views, projection, representation of elements in a design, dimension, etc. are the same. Yet, drawing boards, triangles, scales and other traditional drafting equipment are no longer required to communicate a design idea. A designer/drafter using a computer system and the appropriate software can do the following works.

(1) Plan a part structure, or whatever product is needed.

(2) Modify the design without having to redraw the entire plan.

(3) Call up symbols or base drawings from computer storage.

(4) Automatically duplicate forms and shapes commonly used.

(5) Produce schedules or analyses.

(6) Produce hardcopies of complete drawings or drawing elements in a matter of minutes.

11.3 Building and Information Modeling

Building and information modeling (BIM) is a process involving the generation and management of digital representations of physical and functional characteristics of places. Building information models are files (often but not always in proprietary formats and containing proprietary data) which can be extracted, exchanged or networked to support decision-making regarding a building or other built asset. Current BIM software is used by individuals, businesses and government agencies who plan, design, construct, operate and maintain diverse physical infrastructures, such as water, refuse, electricity, gas, communication utilities, roads, bridges, ports, tunnels, etc.

11.3.1 Development of BIM

Traditional building design was largely reliant upon 2D technical drawings (plans, elevations, sections, etc.). Building information modeling extends this beyond 3D, augmenting the three primary spatial dimensions (width, height and depth) with time as the fourth dimension (4D) and cost as the fifth (5D). BIM therefore covers more than just geometry. It also covers spatial relationships, light analysis, geographic information, and quantities and properties of building components (for example, manufacturers' details).

11.3.2 Function of BIM

BIM involves representing a design as combinations of "objects"—vague and undefined, generic or product-specific, solid shapes or void-space oriented (like the shape of a room), that carry their geometry, relations and attributes. BIM design tools allow extraction of different views from a building model for drawing production and other uses. These different views are automatically consistent, being based on a single definition of each object instance.

BIM software also defines objects parametrically. That is, the objects are defined as parameters and relations to other objects, so that if a related object is amended, dependent ones will automatically also change. Each model element can carry attributes for selecting and ordering them automatically, providing cost estimates as well as material tracking and ordering.

For the professionals involved in a project, BIM enables a virtual information model to be handed from the design team (architects, landscape architects, surveyors, structural and building services engineers, etc.) to the main contractor and subcontractors and then on to the owner/operator. Each professional adds discipline specific data to the single shared model. This reduces information losses that traditionally occurred when a new team takes "ownership" of the project, and provides more extensive information to owners of complex structures.

Words and Phrases

productivity 生产率
earth-moving engineering 土方工程
simulation 模拟
current cost 当时成本,市价
spreadsheet 电子表格
bid-sheet 投标书
payroll 工资单,薪水册
accounting 账目
engineering economy 工程经济
cost control 成本控制
online retrieval 联机检索,在线检索

specification	说明书
code	规范
standard	标准
procurement scheduling	采购计划
inventory control system	库存控制系统
administrative	管理的
statistical analysis	统计分析
acronym	缩写
software package	软件包
record	记录
drawing board	图板
redraw	撤回
building and information modeling	建筑和信息建模
proprietary format	专用格式
spatial relationship	空间关系
amend	修正
subcontractor	分包商

Lesson 12 Hydraulic Engineering

12.1 Introduction

The flow of water in the natural environment, such as rainfall and subsequent infiltration, evaporation, flow in rills and streams, etc., is the purview of surface water hydrology. Because of the uncertainty of natural events, the analysis of hydrologic data requires the use of statistics such as frequency analysis. Hydrology is generally separated into surface water hydrology and subsurface hydrology, depending on whether the emphasis is on surface water or on groundwater. The groundwater engineering is concerned with hydraulics of wells, land subsidence due to excessive pumping, contaminant transport, site remediation and landfills.

Many hydraulic structures have been developed for the storage, conveyance and control of natural flows. These structures include dams, spillways, pipes, open channels, outlet works, energy-dissipating structures, turbines, pumps, etc. The interface between land, ocean and lakes is part of coastal engineering. It also contains a discussion of the mechanics of ocean waves and their transformation in shallow water and resultant coastal circulation. It also includes a discussion of coastal processes and their influence on coastal structures.

12.2 Surface Water Hydrology

Water evaporates from the oceans and land, and becomes a part of the atmosphere. The water vapor is either carried in the atmosphere or it returns to the earth in the form of precipitation. A portion of precipitation falling on land may be intercepted by vegetation and returned back directly to the atmosphere as evaporation. Precipitation that reaches the earth may evaporate or be transpired by plants; or it may flow over the ground surface and reach streams as surface water; or it may infiltrate the soil. The infiltrated water may flow over the upper soil regions and reach surface water or it may percolate into deeper zones and become groundwater. Groundwater may reach the streams naturally or it may be pumped, used and discarded to become a part of the surface water system.

Water on the soil surface enters the soil by infiltration. Percolation is the process by which water moves through the soil because of gravity. As the soil exposed to atmosphere is not usually saturated, flow near the ground surface is through unsaturated medium. The percolated water may reach the ground water storage or it may transpire back to the surface.

An area that drains into a stream at a given location via a network of streams is called a watershed. Rainfall that falls on a watershed fills the depression storage, which consists of storage provided by natural depressions in the landscape, it is temporarily stored on vegetation as interception and it infiltrates into the soil. After these demands are satisfied, water starts flowing over the land and this process is called overland flow. Water that is stored in the upper soil layer may emerge from the soil and join the overland flow. The overland flow lasts only for short distances after which it is collected in small channels called rills. Flows from these rills reach channels. Flow in channels reaches the mainstream.

Important hydrologic processes such as floods or droughts, which are extreme events, are treated as random events. The theory of probability is used to estimate the probabilities of occurrence of these events. The emphasis in statistical analysis is on events rather than on the physical processes that generate them. In the frequency analysis of floods, the emphasis is on the frequency of occurrence of these events.

12.3 Groundwater Engineering

Groundwater engineering is concerned with the occurrence, movement, use and quality of water below ground. The section on fundamentals deals with the definitions, the properties of the unsaturated and saturated zones, and the physics of the movement of subsurface water.

The water table is the level at which the groundwater is at atmospheric pressure. The zone between the ground surface and the water table is called the vadose zone. It contains some water that is held between the soil particles by capillary forces. Immediately above the water table is the capillary fringe where the water fills the pores. The zone above the capillary fringe is often called the unsaturated zone. Below the capillary fringe is the saturated zone. The saturation ratio is the fraction of the volume of voids occupied by water. The water above the water table is below atmospheric pressure while the water below the water table is above atmospheric pressure. Only the water below the water table, the groundwater, is available to supply wells and springs. Recharge of the groundwater occurs primarily by percolation through the unsaturated zone. The geologic formations that yield water in usable quantities, to a well or a spring, are called aquifers. If the upper surface of the saturated zone in the aquifer is free to rise or to decline, the aquifer is said to be an unconfined aquifer. The upper boundary at atmospheric pressure is the water table, also called the phreatic surface. If the water completely fills the formation, the aquifer is confined and the saturated zone is the thickness of the aquifer. If the confining material is impermeable, it is called an aquiclude. If the confining layer is somewhat permeable in the vertical direction, thus permitting slow recharge, it is called an aquitard. When a layer restricts downward infiltration towards the main water table, a perched aquifer with a separate perched water table may be formed. A perched aquifer is, in general, of limited areal extent, and if used as a water supply, extreme caution should be exerted because of its ephemeral nature. If the water in a well in a confined aquifer rises above the top of the aquifer, the water in the aquifer is under pressure, the well is called an artesian well, and the aquifer is in artesian condition. The potentiometric surface, also called the piezometric surface, is defined as the surface connecting the levels to which water will rise in several wells.

Groundwater pumping causes a downward movement of the water table or of the piezometric surface which in turn can cause a downward movement of the land surface called settlement or consolidation. This movement can be a few centimeters to several meters. If the subsidence is not uniform, the differential settlement can produce severe damage to structures.

The management of groundwater requires the capability of

predicting subsurface flow and transport of solutes either under natural conditions or in response to human activities. These modelsare based on the equations governing the flow of water and solutes. These equations are the conservation of mass, Darcy's equation and the contaminant transport equation. When written in two or three dimensions, these are partial differential equations. The complete solute transport model requires at least two equations: one equation for the flow and one for the solute transport. The velocities are obtained from the flow equation. For advectively dominated transport problems, the equations are hyperbolic partial differential equations.

12.4 Water Resources Planning and Management

Water resources planning and management engineering are concerned with conceptualizing, designing and implementing strategies for delivering water of sufficient quality and quantity to meet societal needs in a cost-effective manner. Alternatives that can be engineered to accomplish these functions include development of new water supplies, regulation of natural sources of water, transfer of water over large distances, and treatment of degraded water so that it can be reused. The challenges for water resources engineers are: to identify the essential characteristics of a given water resources problem; to identify feasible alternatives for resolving the problem; to systematically evaluate all feasible alternatives in terms of the goals and objectives of the decision makers; to present a clear and concise representation of the trade-offs that exist between various alternatives.

Among the largest public investments are those designed to stabilize the flow of water in rivers and streams. A stream that may carry little or no water during a significant portion of the year may experience extremely large (perhaps damaging) flows during peak periods. A storage reservoir may be employed to retain water from these peak flow periods for conservation use during low-flow periods (water supply, low-flow augmentation for environmental protection, irrigation, power production, navigation, recreation, etc.) or to contain peak flows for purposes of reducing downstream flood damage (flood control).

Words and Phrases

rainfall	降雨
infiltration	渗透
surface water	地表水
hydrologic	水文的
frequency	频率
well	井
land subsidence	地表沉降
spillway	溢洪道
outlet work	排水工程
energy-dissipating structure	耗能减震结构
turbine	涡轮
infiltrate	渗入,潜入
saturate	饱和
watershed	流域,分水岭
drought	干旱
water table	水位
vadose zone	渗流带
saturation ratio	饱和度
aquifer	含水层

aquiclude	隔水层
perched aquifer	潜水层
artesian well	自流井
feasible alternative	可行的备选方案
concise	简洁的
peak period	峰值期
low-flow	低流量的
environmental protection	环境保护
power production	发电

Lesson 13 Municipal Engineering

13.1 History of Municipal Engineering

Urban development has followed economic reform in China and a large number of central cities rise and develop, making cities in an increasingly important position.

Municipal administration is closely related to modern cities, and the development of cities has driven the formation and improvement of municipal administration. People living in cities require a better living environment to be satisfied from urban life. For all aspects of cities to function normally, a special management system and management mode is needed. The continuous improvement of urban modernization also requires the synchronous development of municipal facilities.

Modern municipal engineering finds its origins in the 19th century in the United Kingdom, following the Industrial Revolution and the growth of large industrial cities. The threat to urban populations from epidemics of waterborne diseases such as cholera and typhus led to the development of a profession devoted to "sanitary science" that later became "municipal engineering".

13.2 Municipal Engineering Content

Municipal engineering is concerned with municipal infrastructure. This involves specifying, designing, constructing and maintaining streets, sidewalks, water supply networks, sewers, street lighting, municipal solid waste management and disposal, storage depots for various bulk materials used for maintenance and public works, public parks and bicycle paths. In the case of underground utility networks, it may also include the civil portion of the local distribution networks of electrical and telecommunication services. It can also include the optimizing of garbage collection and bus service networks. Some of these disciplines overlap with other civil engineering specialties, however, municipal engineering focuses on the coordination of these infrastructure networks and services, as they are often built simultaneously and managed by the same municipal authority.

This engineering is composed of five parts: urban public transportation, urban water supply and drainage engineering, urban gas and heating systems, urban flood control projects, and urban garbage disposal. As students majoring in water supply and drainage engineering, civil engineering, municipal engineering, and roads and bridges, which are closely related to municipal engineering, it is very necessary to learn and understand the basic knowledge of municipal engineering.

13.3 Characteristic of Chinese Urban Construction

Today, municipal engineering may be confused with urban design or urban planning. Whereas the urbanist or urban planner may design the general layout of streets and public places, the municipal engineer is concerned with the detailed design. For example, in the case of a new street, the urbanist may specify the general layout of the street, including landscaping, surface finishing and urban accessories, but the municipal engineering will prepare the detailed plans and specifications for the roads, sidewalks, municipal services and street lighting. As practiced a century ago, municipal engineering, however, fully embraced the function of urban design and urban planning, even though the terms had yet to be coined.

A major characteristic of Chinese urban construction is that ground utilities/housing are developed rapidly in many cities due to the introduction of modern marketing methods, whereas urban infrastructure develops relatively slowly. This causes problems such as delayed marketization, long construction periods and high construction costs for urban infrastructure. As a result, urban infrastructure construction becomes costly and more difficult. When building a new urban infrastructure, specialist technologies, in terms of constructing in limited spaces, or reducing impact on the surrounding environment, are required. Congested commercial and residential buildings surrounding the construction site have seriously complicated the excavation work.

In addition, during construction, special attention should be paid to existing water and sanitation pipes built underground. Therefore, comprehensive investigation and advanced municipal engineering technologies are required to avoid a repeat of these tragedies.

As discussed above, due to construction without sufficient consideration of sustainable development in the early phases of urbanization, new urban infrastructure construction is becoming extremely difficult. Existing utilities and buildings not only increase construction costs, but also affect plans for new utilities' construction.

13.4 Relation to Environment

On the other hand, urban infrastructure must adapt to environmental change. Global warming has raised the issue of city flooding, which is becoming a big problem for most Chinese cities due to inadequate storm drainage utilities. Development of new stormwater drainage systems is therefore urgently required. Despite heavy rains in some regions, lack of rainfall in other regions brought a remarkable water shortage problem, which is still problematic for many inland cities and one of their most serious problems. In recent years, several water transfer projects have been undertaken, for instance the South-North Water Transfer Project. In addition to the current solutions, alternative methods should be considered. For instance, storm water can be reused by constructing storm water storage infrastructure. Other problems such as environmental protection and energy saving should be taken into consideration in urban development.

Words and Phrases

municipal engineering 市政工程
economic reform 经济改革
municipal administration 市政管理
urban modernization 城市现代化
synchronous development 同步发展
waterborne disease 水传播疾病
epidemics 泛滥
cholera 霍乱
typhus 斑疹伤寒
storage depot 储存库

overlap	重叠
flood control project	防洪工程
sanitation pipe	卫生管道
city flooding	城市洪水
inland city	内陆城市
South-North Water Transfer Project	南水北调工程
alternative method	替代方法
storm water	地表径流
energy saving	节能
urban development	城市发展

Lesson 14 Mechanical Property

14.1 Introduction

The strength of a material is almost always the first property that the engineer needs to know about. If the strength is not adequate, then the material cannot be used and other properties are not even considered. The next property to be considered is often the "stiffness" or elastic modulus, because this determines how far a structure will deflect under load.

Materials for building must have certain physical properties to be structurally useful. Primarily they must be able to carry a load or weight, without changing shape permanently. When a load is applied to a structure member, it will deform, that is, a wire will stretch or a beam will bend. However, when the load is removed, the wire and the beam come back to the original positions. This material property is called elasticity. If a material were not elastic and a deformation were present in the structure after removal of the load, repeated loading and unloading eventually would increase the deformation to the point where the structure would become useless. All materials used in architectural structures, such as stone, brick, wood, steel, aluminum, reinforced concrete and plastics, behave elastically within a certain defined range of loading. If the loading is increased above the range, two types of behavior can occur: brittle and plastic. In the former, the material will break suddenly. In the latter, the material begins to flow at a certain load (yield strength), ultimately leading to fracture. As examples, steel exhibits plastic behavior, and stone is brittle. The ultimate strength of a material is measured by the stress at which failure (fracture) occurs.

A second important property of a building material is its stiffness. This property is defined by the elastic modulus, which is the ratio of the stress (force per unit area) to the strain (deformation per unit length). The elastic modulus, therefore, is a measure of the resistance of a material to deformation under load. For two materials of equal area under the same load, the one with the higher elastic modulus has the smaller deformation. Structural steel, which has an elastic modulus of 30×10^6 lb/in^2 (psi, 1 psi = 6.895 kPa), or 2,100,000 kg/cm^2, is 3 times as stiff as aluminum, 10 times as stiff as concrete, and 15 times as stiff as wood.

14.2 Mass and Gravity

In the MKS SI system, the mass of an object is defined from its acceleration when a force is applied, for example, we have

$$f = ma \tag{14.1}$$

where f is the force in N; m is the mass in kg; a is the acceleration in m/s^2.

Gravity is normally the largest force acting on a structure. On the earth's surface, the gravitational force on a mass m is given by

$$f = mg \tag{14.2}$$

where g is the gravitational constant, $g = 9.81\ m/s^2$.

The gravitational force on an object is called its weight. Thus, an object will have a weight of 9.81 N/kg of mass. An approximate value of 10 is often used for g to give the commonly used value of 10 kN weight for a mass of 1 t. In the US customary system of units, force is generally measured as a weight in pounds and, if this is done, a constant term for $g = 32.2\ ft/s^2$ must be included in Equation (14.1).

14.3 Stress and Strain

14.3.1 Type of strength

In engineering, the term strength is always defined by type, and is probably one of the following (Figure 14.1), depending on the method of loading: compressive strength, tensile strength and flexural strength.

14.3.2 Tensile and compressive stress

In order to define strength, it is necessary to define stress. This is a measure of the internal resistance in a material to an externally applied load. For direct compressive or tensile loading, the stress is designated σ, and is defined in

$$\text{Stress}, \ \sigma = \frac{\text{Load}(W)}{\text{Area}(A)} \tag{14.3}$$

and measured in N/m^2 or psi.

Load and stress is shown in Figure 14.2.

14.3.3 Strain

In engineering, strain is not a measure of force, but is a measure

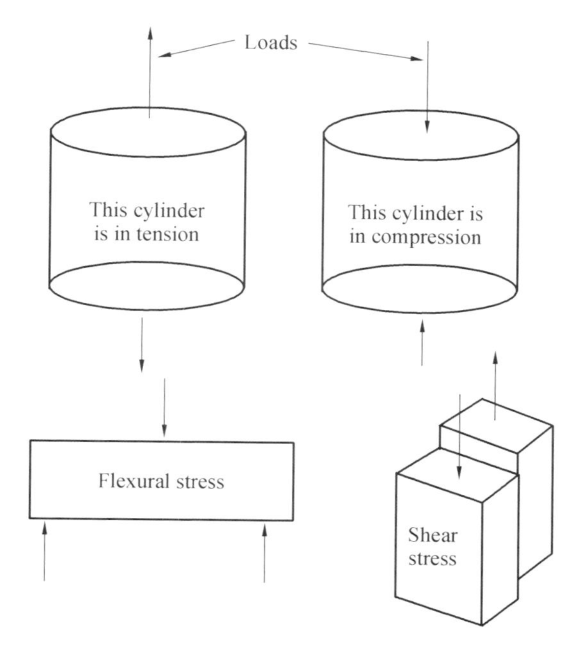

Figure 14.1 Compression, tension, flexure and shear

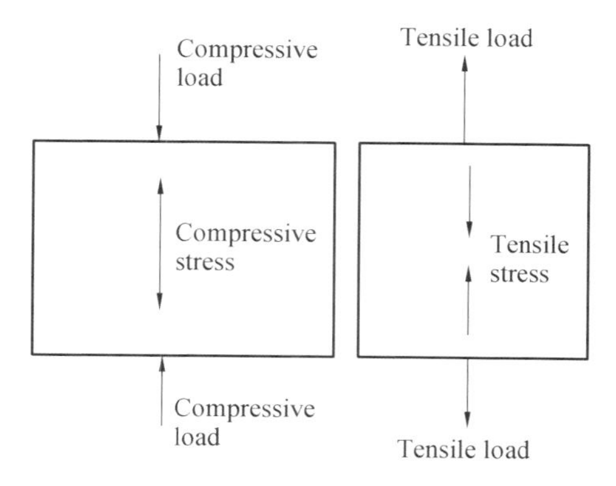

Figure 14.2 Load and stress

of the deformation produced by the influence of stress. For tensile and compressive loads, strain is dimensionless, so it is not measured in m, kg, etc. The commonly used unit is microstrain (mstrain), which is a strain of one part per million, so we have

$$\text{Strain} = \frac{\text{Increase}(x)}{\text{Original}(L)} \tag{14.4}$$

For shear loads, the strain is defined as the angle (Figure 14.3). This is measured in radians, and thus for small strains.

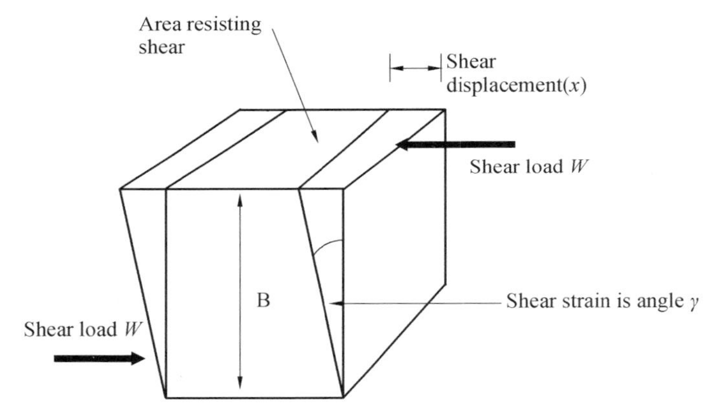

Figure 14.3 Shear stress and strain

The arrows show the sequence of loading and unloading during a test.

14.3.4 Deformation and strength

Strain may be elastic or plastic. The stress on an object, and the resulting strain as it is loaded and then unloaded. If the strain is elastic, the sample returns exactly to its initial shape when unloaded. If plastic strain occurs, there is permanent deformation. If the material exhibits plastic deformation (yields), and does not return to its original shape when unloaded, this is clearly unacceptable for most construction applications. Figure 14.4 shows a stress-strain curve for a typical metal. As the load is applied, the graph is initially linear (the stress is proportional to the strain), until it reaches a yield point. If the load is removed after yield, the sample will not return to its original shape, and is left with final residual strain.

For a brittle material (such as concrete), strength is defined from the stress at fracture, but for a ductile material (e. g. some steels) that yields a long way before failure, strength is often defined from limits to the residual strain, after loading and unloading.

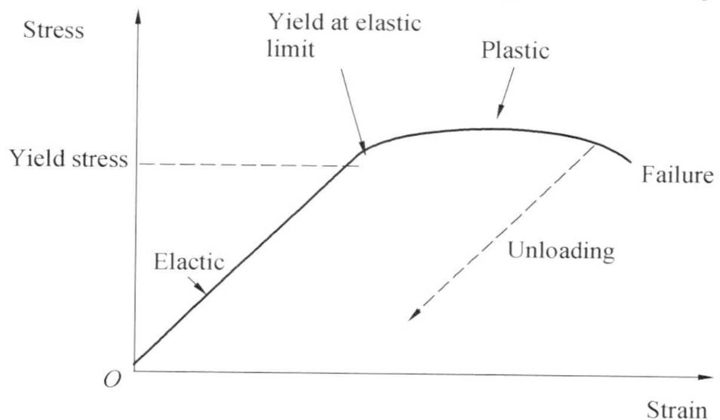

Figure 14.4 Elastic and plastic strain

14.3.5 Modulus of elasticity

If the strain is "elastic", that is, on the linear part of a graph of stress versus strain, Hooke's law may be used to define Young's modulus as the gradient, that is

$$E=\frac{\text{Stress}}{\text{Strain}} \tag{14.5}$$

Thus we have

$$E=\frac{W}{x}\times\frac{L}{A} \tag{14.6}$$

where $\frac{W}{x}$ may be the gradient of a graph of load versus displacement obtained from an experiment. The Young's modulus is also called the modulus of elasticity or stiffness, and is a measure of how much strain occurs due to a given stress. Because strain is dimensionless, Young's modulus has the units of stress or pressure.

In reality, no part of a stress – strain curve obtained from an experiment is ever perfectly linear. Thus the modulus must be obtained from a tangent or a secant. The difference between an initial tangent and secant modulus is shown in Figure 14.5.

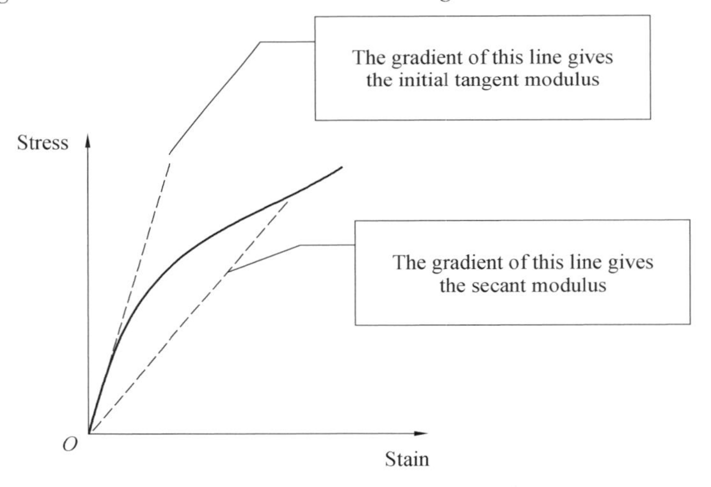

Figure 14.5 Tangent and secant modulus

14.3.6 Poisson's ration

This is a measure of the amount by which a solid "spreads out sideways" under the action of a load from above. It is defined from

$$\text{Poisson'sration} = \frac{\text{Lateralstrain}}{\text{Strain in directionload}} \tag{14.7}$$

A material like timber which has a "grain direction" will have a number of different Poisson's ratios corresponding to loading, and deformation in different directions.

14.3.7 Fatigue strength

If a material is continually loaded and unloaded (e. g. the springs in a car), the permanent strain from each cycle slowly decreases. This may be seen from Figure 14.6. Eventually, the sample will fail, and the number of cycles it takes to fail will depend on the maximum stress that is being applied.

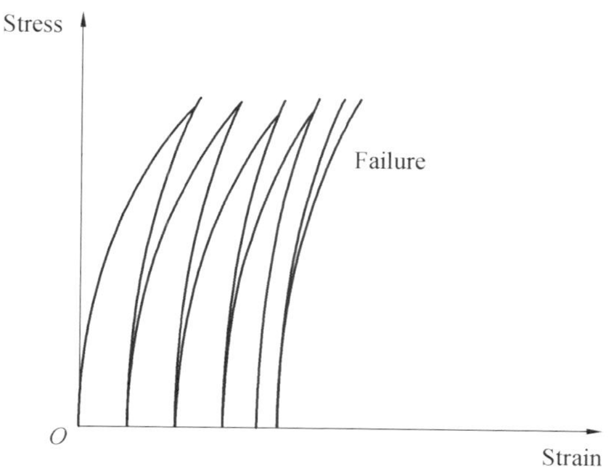

Figure 14.6 Fatigue cycles to failure

14.4 Building Mechanics

"Building mechanics" is about the working condition of the structures and components affected by the loads or other factors, which is the strength, stiffness and stability of buildings. It is a very wide range of areas, including theoretical mechanics, material mechanics, structural mechanics, elastic mechanics, plastic mechanics, structural dynamics, etc. We will introduce three main machanics in the followings.

14.4.1 Theoretical mechanics

It is a discipline that studies general rules of the mechanical movement among or between objects. The content includes statics, kinematics and dynamics. By theoretical mechanics, we can abstract the simple practical problems in practice work to the static model to analyze the balance and the force of the objects; to carry out the calculation about general force systems; to analyze a simple movement about a point or a rigid body.

14.4.2 Material mechanics

It is a research about stress, strain, strength and stiffness in various forces, and the limits when all kinds of materials damage.

The main task of material mechanics is to study the destruction law of materials acted forces on, which may be steel, concrete, wood, plastics, etc., and the requested strength, stiffness and stability of bars, which may be beams, columns and other components, whose length is more than two horizontal sizes, to solve the contradiction between structural reliability and economic reasonableness.

Through material mechanics courses, students should be able to calculate the stiffness and the strength, when simple deformation comes on the straight bar in pulling, pushing, cutting, twisting, bending; to calculate the stress and strength, when curved ramps, eccentric compression (tension) bending and torsion comes on the combined deformation; to understand the purposes and ways of the basic experiments.

14.4.3 Structural mechanics

It is a discipline to study the law of structures bearing and transferring loads and how to optimize the structures. Here, the structure means the system that can bear and transfer loads, including rods, plates, shells and their combinations, such as bridges, roof trusses and load, bearing walls, etc.

The main task of structural mechanics courses is to study the laws of stress, strain and displacement, when the loads are acted on the structures; to analyze engineering structures, which are made up of different materials or forms, to provide analytic methods and calculable formulas for engineering design; to identify the ability of bearing and transferring loads; to research and develop new types of engineering structures.

Structural mechanics isn't only an ancient discipline, but also a rapidly developing discipline. Numbers of new materials and new structures coming on can provide a new research and put forward new demands. The development of computer may provide a powerful tool for structural mechanics. Structural mechanics also plays a role in promoting the development of mathematics and other disciplines.

Words and Phrases

stiffness 刚度

elastic modulus 弹性模量

deform 变形

bend 弯曲

original position 原点

unload 卸载

useless 失效

brittle 脆性

plastic 塑性

yield strength	屈服强度
ultimate strength	极限强度
MKS SI system	米-千克-秒国际体系
mass	质量
acceleration	加速度
flexural strength	抗弯强度
strain	应变
initial shape	原形
permanent deformation	永久变形
ductile material	延展性材料
Hooke's law	胡克定律
Young's modulus	杨氏模量
gradient	斜率
dimensionless	无量纲
initial tangent	初始切线
secant modulus	割线模量
fatigue strength	疲劳强度
building mechanics	建筑力学
theoretical mechanics	理论力学
material mechanics	材料力学
structural mechanics	结构力学

Lesson 15 Loads in Structure

15.1 Development of Definition

The primary objective of a course in materials mechanics is the development of relationships between the loads applied to a non-rigid body and the internal forces and deformations induced in the body. Ever since the time of Galileo Galilei (1564–1642), men of scientific bent have studied the problem of the load-carrying capacity of structural members and machine components, and have developed mathematical and experimental methods of analysis for determining the internal forces and the deformations induced by the applied loads. The experiences and observations of these scientists and engineers of the last three centuries are the heritage of the engineer of today. The fundamental knowledge gained over the last three centuries, together with the theories and analysis techniques developed, permit the modern engineer to design, with complete competence and assurance, structures and machines of unprecedented size and complexity. Certain terms are commonly used to describe applied loads, their definitions are given here so that the terminology will be clearly understood.

15.2 Classification by Variation in Time

In general, loads that act on building structures can be divided into three groups (dead load, live load and accidental load), because of gravitational attraction and those resulting from other natural causes and elements.

15.2.1 Dead load

The load is permanently applied to the structure during use and its value does not change with time, also known as permanent load. The weight of the load-bearing structure, the gravity of the envelope structure (wall, floor, roof, bridge floor, etc.) and the gravity of the fixed device are the dead load. Dead load is generally obtained by multiplying the volume of the member by the mass per unit volume of the material used. In buildings, the dead load of load-bearing structures accounts 50% –70% of the total load. In addition, the soil pressure and surrounding rock pressure of buried underground, retaining soil and tunnel engineering facilities are also dead loads.

15.2.2 Live load

Load applied to the structure during the use of the value changes with time, also known as variable load. There are some types of live load.

1. Service live load

Floor live load (people, furniture, mobile equipment, work pieces used in operation, etc.), roof live load (people and facilities on the roof, roof ash accumulation, roof helipad, etc.) are all used.

2. Vehicle live load

The vehicle live load is mainly the load of railway trains or highway vehicles, which is manifested as a series of moving loads composed of concentrated loads and uniform loads. In addition, the lifting crane in the industrial plant (commonly known as the crane) is also a mobile live load.

3. Wind load

Wind is the movement of air caused by unequal atmospheric pressure. The pressure exerted on the wall or structure by the movement of the wind is called wind load. There are three "differences" in the wind during designing: different regions are different (large coastal, small inland); the same area is different from time to time; different elevations and different parts are different (the wind load at 50 m is about 1.5 times greater than that at 5 m altitude). The wind load value used in the design is the statistical value of the average wind pressure in 10 min with a recurrence period of 50 years.

4. Snow load

Snow load refers to the load caused by snow. Snow load also varies from region to region. The southern parts of China, such as Guangzhou and Haikou, even do not have snow.

15.2.3 Accidental load

The load does not necessarily appear during use, but once it appears, its value is very large and the duration is very short, such as impact load, explosion load, etc.

15.3 Classification of Loads with Structural Response

A distinction is made between two types of load according to the response of the structure.

15.3.1 Static load

Static load is applied to the structure without causing significant accelerations of the structure or of structural elements.

15.3.2 Dynamic load

Dynamic load causes significant accelerations of the structure. The same load might be static or dynamic depending upon the structure to which it is applied. Generally, loads can be considered static loads, provided that the dynamic effects are taken into account by an increment of the intensity of the loads. In other cases, a dynamic analysis is necessary.

15.4 Conclusion

One final type of load is an impact load, usually due to moving equipment, which occurs within or on the structure. Most structural materials can withstand a sudden and temporary load of higher magnitudes are substantially increased when such loads govern the design. No permanent damage is done by a moderate impact load provided that it does not occur repeatedly. (An earthquake is a good example of a severe and repeating impact load.)

The designer should consult local building codes, which always take precedence. The designer also bears the professional responsibility for increasing any recommended design loads when the situation warrants is changed.

Words and Phrases

non-rigid body	非刚性体
internal force	内力
time of Galileo Galilei	伽利略时代
scientific bent	科学追求
load-carrying capacity	承载能力
machine component	机械零件
fundamental knowledge	基础知识
unprecedented size	规模空前的
applied load	施加的荷载
terminology	术语
variation	变化,变量
dead load	恒载
live load	活载
accidental load	偶然荷载
permanent load	永久荷载
bridge floor	桥面
soil pressure	土压力
surrounding rock pressure	围岩压力
variable load	可变荷载
service live load	使用活载
vehicle live load	车辆活载
lifting crane	吊车
duration	持续时间
static load	静载
dynamic load	动载
local building code	当地建筑规范
design load	设计荷载

Lesson 16 Principles of Static

Statics consists of the study of structures that are at rest under equilibrium conditions. To ensure equilibrium, the forces acting on a structure must balance, and there must be no net torque acting on the structure. The principles of statics provide the means to analyze and determine the internal and external forces acting on a structure.

For planar structures, three equations of equilibrium are available for the determination of external and internal forces. A statically determinate structure is one in which all the unknown member forces and external reactions may be determined by applying the equations of equilibrium.

An indeterminate or redundant structure is one that possesses more unknown member forces or reactions than the available equations of equilibrium. These additional forces or reactions are termed redundants. To determine the redundants, additional equations must be obtained from conditions of geometrical compatibility. The redundants may be removed from the structure, and a stable, determinate structure remains, which is known as the cut-back structure. External redundants are that exist among the external reactions. Internal redundants are that exist among the member forces.

16.1 Representation of Forces

A force is an action that tends to maintain or change the position of a structure. The forces acting on a structure are the applied loads, consisting of both dead and imposed loads, and support reactions. As shown in Figure 16.1, the simply supported beam is loaded with an imposed load W_{LL} located at point 3 and with its own weight w_{DL}, which is uniformly distributed over the length of the beam. The support reactions consist of the two vertical forces located at the ends of the beam. The lines of application of all forces on the beam are parallel.

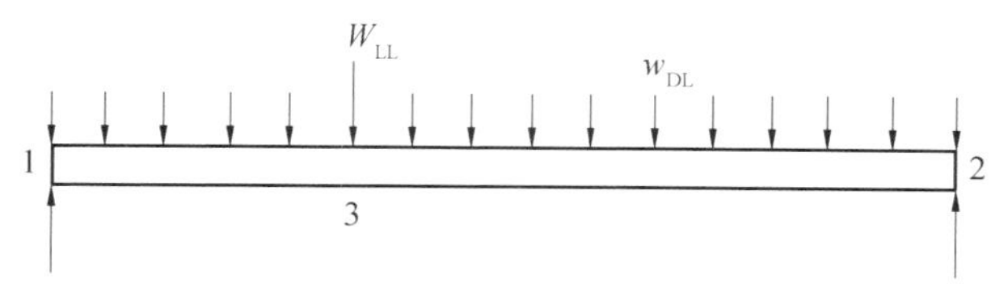

Figure 16.1 Simply supported beam

In general, a force may be represented by a vector quantity having a magnitude, location and direction corresponding to the force. A vector represents a force to scale, such as a line segment with the same line of action as the force and with an arrowhead to indicate direction.

The point of application of a force along its line of action does not affect the equilibrium of a structure. However, as shown in the three-hinged portal frame in Figure 16.2, changing the point of application may alter the internal forces in the individual members of the structure.

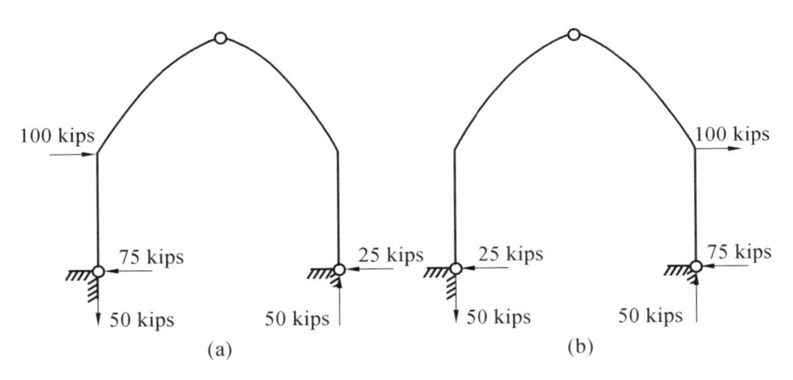

Figure 16.2 Three-hinged portal frame

Collinear forces are forces acting along the same line of action. The two horizontal forces acting on the portal frame shown in Figure 16.3 (a) are collinear and may be added to give the single resultant force shown in Figure 16.3(b).

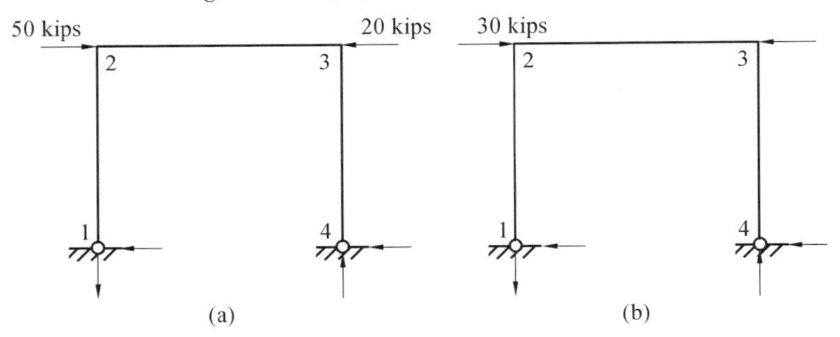

Figure 16.3 Portal frame

Forces acting in one plane are coplanar forces. Space structures are three-dimensional structures and, as shown in Figure 16.4, may be acted on by noncoplanar forces.

In a concurrent force system, the line of action of all forces has a common point of intersection. As shown in Figure 16.5, for equilibrium of the two-hinged arch, the two reactions and the applied load are concurrent.

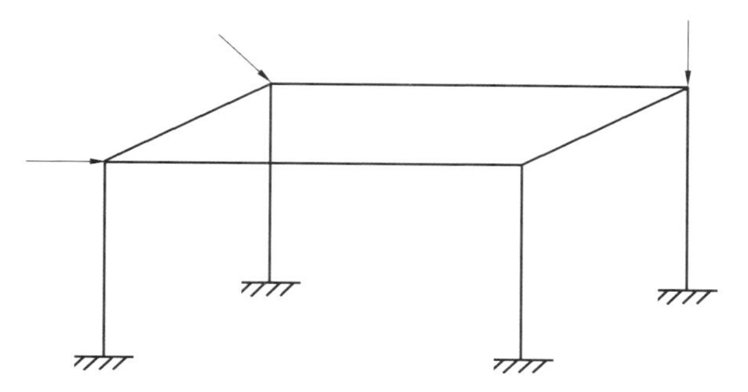

Figure 16.4 Space structures

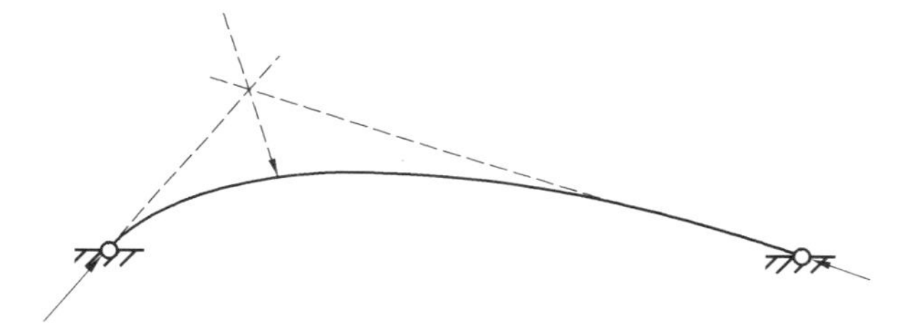

Figure 16.5 Two-hinged arch

It is often convenient to resolve a force into two concurrent components. The original force then represents the resultant of the two components. The directions adopted for the resolved forces are typically the x- and y-components in a rectangular coordinate system. As shown in Figure 16.6, the applied force F on the arch is resolved into the two rectangular components:

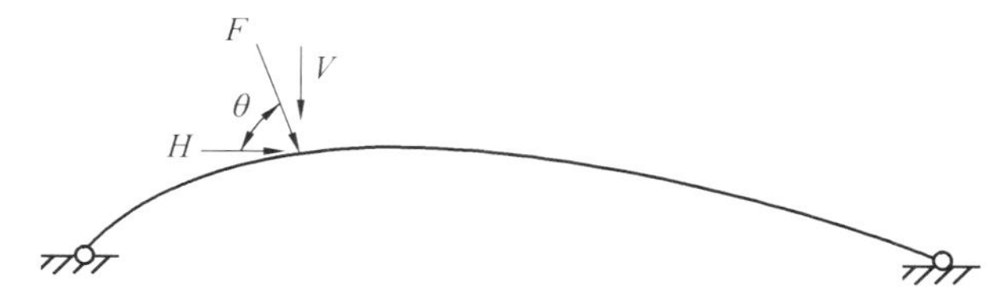

Figure 16.6 x- and y-components

$$H = F\cos\theta$$
$$V = F\sin\theta$$

16.2 Conditions of Equilibrium

In order to apply the principles of statics to a structural system, the structure must be at rest. This is achieved when the sum of the applied loads and support reactions is zero and there is no resultant couple at any point in the structure. For this situation, all component parts of the structural system are also in equilibrium.

A structure is in equilibrium with a system of applied loads when the resultand force in any direction and the resultant moment about any point is zero.

For a system of coplanar forces, this may be expressed by the three equations of static equilibrium:

$$\sum H = 0$$
$$\sum V = 0$$
$$\sum M = 0$$

where H and V are the resolved components in the horizontal and vertical directions of a force; M is the moment of a force about any point.

16.3 Sign Convention

For a planar, two-dimensional structure subjected to forces acting in the xy plane, the sign convention adopted is shown in Figure 16.7. Using the righthand system as indicated, horizontal forces acting to the right are positive and vertical forces acting upward are positive. The z-axis points out of the plane of the paper, and the positive direction of a couple is that of a right-hand screw progressing in the direction of *the* z-axis. Hence, counterclockwise moments are positive.

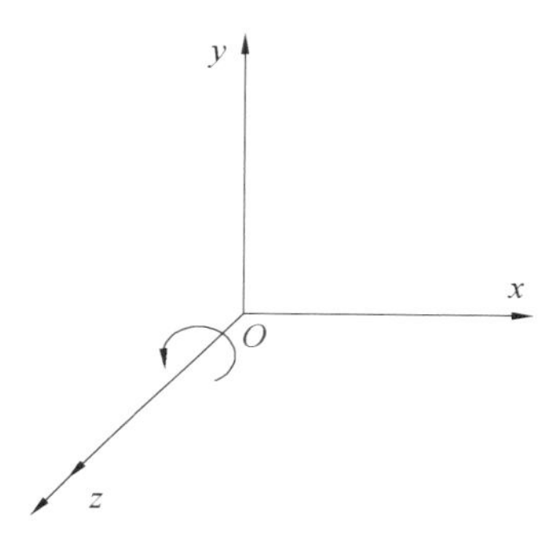

Figure 16.7 Righthand system

16.4 Triangle of Forces

When a structure is in equilibrium under the action of three concurrent forces, the forces form a triangle of forces. As indicated in Figure 16.8 (a), the three forces F_1, F_2 and F_3 are concurrent. As shown in Figure 16.8 (b), if the initial point of force vector F_2 is placed at the terminal point of force vector F_1, then the force vector F_3 drawn from the terminal point of force vector F_2 to the initial point of force vector F_1 is the equilibrant of F_1 and F_2. Similarly, as shown

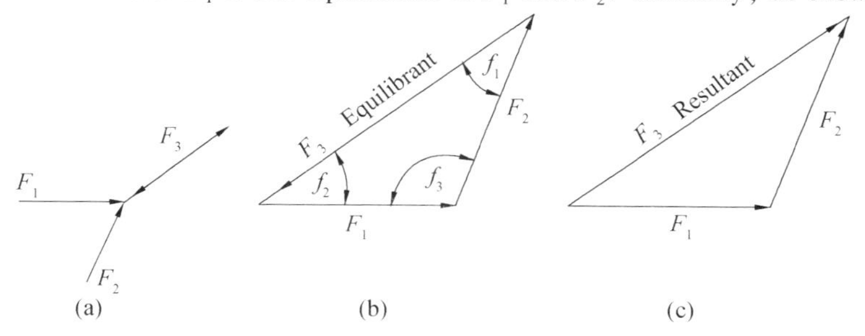

Figure 16.8 Concurrent forces

in Figure 16. 8 (c), if the force vector F_3 is drawn from the initial point of force vector F_1 to the terminal point of force vector F_2, this is the resultant of F_1 and F_2. The magnitude of the resultant is given algebraically by

$$F_3^2 = F_1^2 + F_2^2 - 2F_1F_2\cos f_3 \tag{16.1}$$

16.5 Free Body Diagram

For a system in equilibrium, all component parts of the system must also be in equilibrium. This provides a convenient means for determining the internal forces in a structure using the concept of a free body diagram. Figure 16. 9 (a) shows the applied loads and support reactions acting on the pin-jointed truss that was analyzed in example. The structure is cut at section A–A, and the two parts of the truss are separated as shown at Figure 16.9(b) and (c) to form two free body diagrams. The left-hand portion of the truss is in equilibrium under the actions of the support reactions of the complete structure at joint 1, the applied loads at joint 3, and the internal forces acting on it from the right-hand portion of the structure. Similarly, the right-hand portion of the truss is in equilibrium under the actions of the support reactions of the complete structure at joint 2, the applied load at joint 4, and the internal forces acting on it from the lefthand portion of the structure. The internal forces in the members consist of a compressive force in member 34 and a tensile force in members 45 and 25. By using the three equations of equilibrium on either of the free body diagrams, the internal forces in the members at the cut line may be obtained. The values of the member forces are indicated at Figure 16.9(b) and (c).

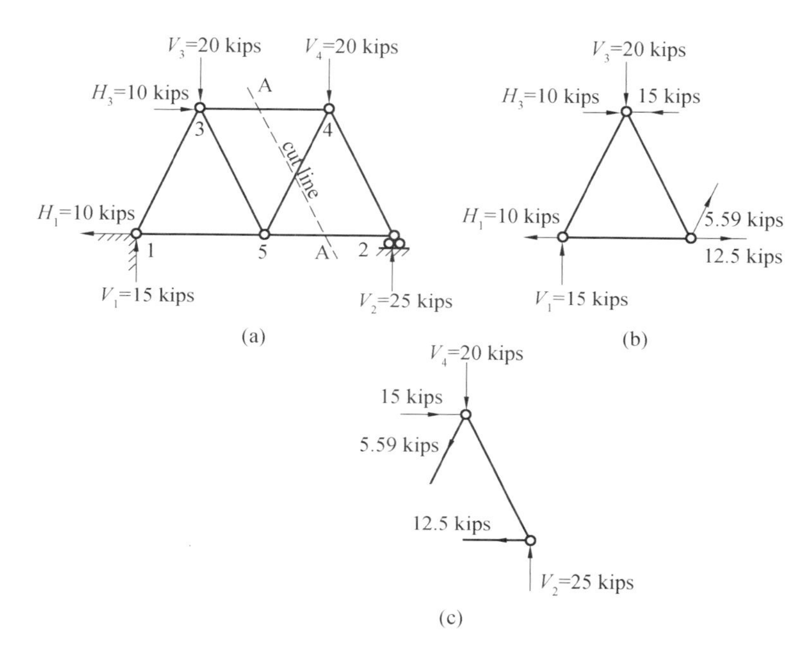

Figure 16.9 Example for cut section

16.6 Principle of Superposition

The principle of superposition may be defined as follows: the total displacements and internal tresses in a linear structure corresponding to a system of applied forces is the sum of thedisplacements and stresses corresponding to each force applied separately. The principle applies to all linear-elastic structures in which displacements are proportional to applied loads and which are constructed from materials with a linear stress – strain relationship. This enables loading on a structure to be broken down into simpler components to facilitate analysis. As shown in Figure 16. 10, a pin-jointed truss is subjected to two vertical loads in Figure 16. 10 (a) and a horizontal load in Figure 16.10 (b). The support reactions for each loading case are shown. As shown in Figure 16.10(c), the principle of superposition and the two loading cases may be applied simultaneously to the truss, producing the combined support reactions indicated.

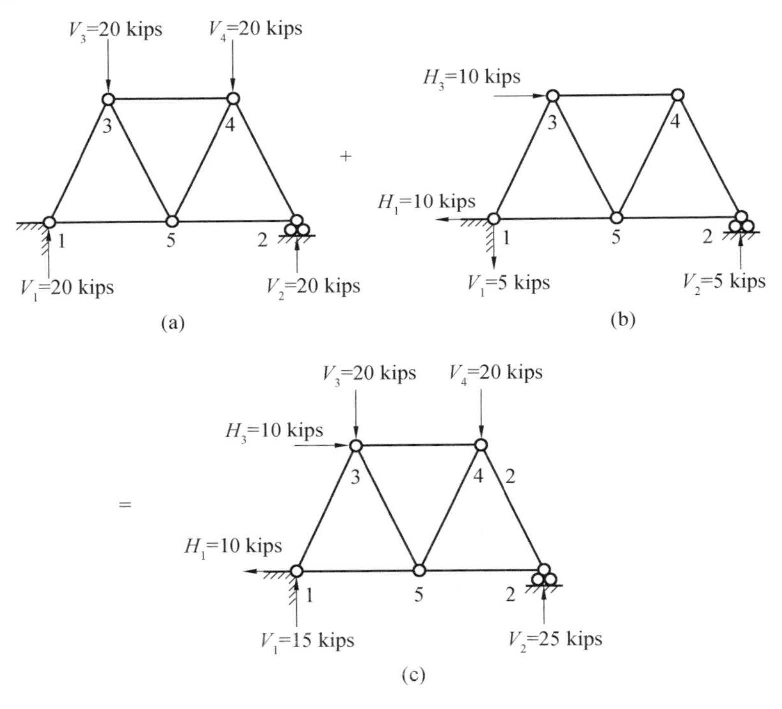

Figure 16.10 Example for principle of superposition

Words and Phrases

principles of static	静定原理
equilibrium conditions	平衡条件
net torque	净扭矩

planar structure	平面结构
equation of equilibrium	平衡方程
statically determinate structure	静定结构
redundant structure	超静定(静不定)结构
reaction	反力
redundant	超静定(多余)杆件
geometrical	几何的
imposed load	施加的荷载
uniformly distributed	均布
support reaction	支座反力
line of application	(力的)作用线
vector quantity	向量
point of application	作用点
collinear force	共线力
coplanar force	共面力
concurrent force system	汇交力系
two-hinged arch	两铰拱
concurrent	共点的
condition of equilibrium	平衡条件
moment	力矩
sign convention	符号规定
positive direction	正向
counterclockwise	逆时针方向
terminal point	终点
free body diagram	隔离体受力图
pin-jointed	铰接的

section	截面
principle of superposition	叠加原理

Lesson 17 Construction Operations

17.1 Construction

Construction operations are generally classified according to specialized fields. These include preparation of the project site, earthmoving, foundation treatment, steel erection, concrete placement, asphalt paving, and electrical and mechanical installations. Procedures for each of these fields are generally the same, even when applied to different projects, such as buildings, dams or airports. However, the relative importance of each field is not the same in all cases.

17.2 Preparation of Site

This consists of the removal and clearing of all surface structures and growth from the site of the proposed structure. A bulldozer is used for small structures and trees. Larger structures must be dismantled.

17.3 Earthmoving

This includes excavation and the placement of earth fill.

Excavation follows preparation of the site, and is performed when the existing grade must be brought down to a new elevation. Excavation generally starts with the separate stripping of the organic topsoil, which is later reused for landscaping around the new structure. This also prevents contamination of the nonorganic material which is below the topsoil and which may be required for fill. Excavation may be done by any of several excavators, such as shovels, draglines, clamshells, cranes and scrapers.

Efficient excavation on land requires a dry excavation area, because many soils are unstable when wet and cannot support excavating and hauling equipment. Dewatering becomes a major operation when the excavation lies to low the natural water table and intercepts the groundwater flow. When this occurs, dewatering and stabilizing of the soil may be accomplished by trenches, which conduct seepage to a sump from which the water is pumped out. Dewatering and stabilizing of the soil may in other cases be accomplished by wellpoints and electroosmosis.

Some materials, such as rock, cemented gravels and hard clay, require blasting to loosen or fragment the material. Blast holes are drilled in the material, explosives are then placed in the blast holes and detonated. The quantity of explosives and the blast-hole spacing are dependent upon the type and structure of the rock, and the diameter and depth of the blast holes.

After placement of the earth fill, it is almost always compacted to prevent subsequent settlement. Compaction is generally done with sheep-foot, pneumatic-tired and vibratory-type rollers, which are towed by tractors over the fill as it is being placed. Hand-held, gasoline-driven hammers are used for compaction close to structures where there is no room for rollers to operate.

17.4 Foundation Treatment

When subsurface investigation reveals structural defects in the foundation area which to be used for a structure, the foundation must be strengthened. Water passages, cavities, fissures, faults and other defects are filled and strengthened by grouting. Grouting consists of injection of fluid mixtures under pressure. The fluids subsequently solidify in the voids of the strata. Most grouting is done with cement and water mixtures, but other mixture ingredients are asphalt, cement and clay, and precipitating chemicals.

17.5 Steel Erection

The construction of a steel structure consists of the assembly at the site of mill-rolled or shop-fabricated steel sections. The steel sections may consist of beams, columns or small trusses which are joined together by riveting, bolting or welding. It is more economical to assemble sections of the structure at a fabricating shop rather than in the field, but the size of preassembled units is limited by the capacity of transportation and erection equipment. The crane is the most common type of erection equipment, but when a structure is too high or extensive in area to be erected by a crane, it is necessary to place one or more derricks on the structure to handle the steel. In high structures, the derrick must be constantly dismantled and reerected to successively higher levels to raise the structure. For river bridges, the steel may be handled by cranes on barges, or, if the bridge is too high, by traveling derricks which ride on the bridge being erected. Cables for long suspension bridges are assembled in place by special equipment that pulls the wire from a reel, set up at one anchorage, across to the opposite anchorage, repeating the operation until the bundle of wires is of the required size.

17.6 Concrete Construction

Concrete construction consists of several operations: forming, concrete production, placement and concrete curing. Forming is required to contain and support the fluid concrete within its desired final outline until it solidifies and can support itself. The form is made of timber or steel sections or a combination of both and is held together during the concrete placing by external bracing or internal ties. The forms and ties are designed to withstand the temporary fluid pressure of the concrete.

The usual practice for vertical walls is to leave the forms in position for at least a day after the concrete is placed. They are removed when the concrete has solidified or set. Slip-forming is a method where the form is constantly in motion, just ahead of the level of fresh concrete. The form is lifted upward by means of jacks which are mounted on vertical rods embedded in the concrete and are spaced along the perimeter of the structure. Slip forms are used for high structures such as silos, tanks or chimneys.

Concrete may be obtained from commercial batch plants which deliver it in mix trucks if the job is close to such a plant, or it may be produced at the job site. Concrete production at the job site requires the erection of a mixing plant, and of cement and aggregate receiving and handling plants. Aggregates are sometimes produced at or near the job site. This requires opening a quarry and erecting processing equipment such as crushers and screens.

Concrete is placed by chuting directly from the mix truck, where possible, or from buckets handled by means of cranes or cableways, or it can be pumped into place by special concrete pumps.

Curing of exposed surfaces is required to prevent evaporation of mix water or to replace moisture that does evaporate. The proper balance of water and cement is required to develop full design strength.

Concrete paving for airports and highways is a fully mechanized operation. Batches of concrete are placed between the road forms from a mix truck or a movable paver, which is a combination mixer and placer. A series of specialized pieces of equipment, which ride on the forms, follow to spread and vibrate the concrete, smooth its surface, cut contraction joints, and apply a curing compound.

17.7 Asphalt Paving

This is an amalgam of crushed aggregate and a bituminous binder. It may be placed on the roadbed in separate operations or mixed in a mix plant and spread at one time on the roadbed. Then the pavement is compacted by rollers.

Words and Phrases

earthmoving	土方工程
foundation	基础
steel erection	钢结构安装
concrete placement	浇筑混凝土
asphalt paving	铺设沥青
electrical and mechanical installation	机电安装
site of the proposed structure	拟建场地
excavation	开挖
organic topsoil	表层有机土
excavator	挖掘机

hauling equipment	运输设备
dewater	降水
wellpoint	井点
electroosmosis	电渗
hard clay	硬质黏土
blast hole	炮眼
earth fill	填方土
strengthen	加强,加固
grouting	注浆
preassembled unit	预制构件
derrick	吊车
anchorage	锚碇
forming	支模
concrete curing	混凝土养护
slip-forming	滑模法
commercial batch plant	商混站
job site	施工现场
mix truck	搅拌车
design strength	设计强度
bituminous binder	沥青胶合剂,沥青结合料
spread	摊铺
roller	压路机

Lesson 18 Engineering Tendering and Bidding

18.1 Introduction

Tendering and bidding are widely used as international trading methods for construction projects, with the aim of selecting appropriate contractors. Tendering can be seen as a purchasing method for civil engineering product demanders, while bidding can be as a sales method for civil engineering product producers. The principle of bidding is to encourage competition and prevent monopolies, and the World Bank and other international financial institutions have strict regulations on the procedures and rules of bidding.

According to the Tendering and Bidding Law, bidding is required for purchasing important equipment and materials in such projects as survey, design, construction, supervision in China.

(1) Large scale infrastructure, public utilities and other projects related to social public interests and public safety.

(2) Projects that use all or part of state-owned funds for investment or state financing.

(3) Projects that use loans or aid funds from international

organizations or foreign governments.

Tendering can be conducted in an open and inviting manner. Open tenders refer to inviting unspecified legal persons or other organizations to bid in the form of tender announcements; while the specific legal persons or other organizations are invited to bid in the form of invitation tenders.

Tender and bidding encompasses three main stages: tendering, bidding and awarding. Bidding is a crucial aspect of construction project management.

18.2 Tender

There are three works to do in tender stage: preparing tender documents, issuance tender announcements and the pre-qualification examination.

18.2.1 Tender documents

The bidding documents (tender documents) of engineering construction generally include the following contents: invitation to bid, instructions to bidders, main terms of the contract and the format of bidding documents. Where a bill of quantities is used for bidding, a bill of quantities, technical terms, design drawings, evaluation standards and methods, and auxiliary tender materials should be provided.

18.2.2 Tender announcement

The tender announcement generally includes the name and address of the tenderer, the content, scale and funding sources of the tender project, the implementation place and duration (scheduled commencement and completion date), the place, time and fee to obtain the tender documents, the requirements for the qualification level of the tenderer, the bid bond, etc.

18.2.3 Pre-qualification

Pre-qualification is a written investigation by the tendering unit on the financial strength, technical level, construction management experience and other aspects of many contractors who have signed up to participate in the bid to assess whether the contractor has the ability to complete the contract task. The pre-qualification examination shall be conducted in accordance with the conditions of qualified bidders specified in the bidding notice, and no unreasonable conditions shall restrict or exclude potential bidders.

18.3 Bidding

Bidding includes the investigation of the status of the project and the preparation of tender documents.

18.3.1 Investigate the project status

By bidding documents, inspecting the site, participating in the bidding disclosure meeting and other ways, the situation of the project can be investigated.

18.3.2 Tender documents

Documents include technical bids and business bids. The technical one is composed of project overview, construction plan, construction organization design, the measures to ensure the quality and duration of the project, and how to ensure the construction safety and environmental protection. Business bid is the quotation, according to the list of quantities provided in the tender documents, calculate the project cost. After considering the bidding strategy and various factors affecting the cost of the project, the offer is made.

18.4 Award of Bid

The award stage includes bid opening, evaluation and outbid.

18.4.1 Bid opening

Bid opening will be conducted in public at the same time as the deadline for submission of bid documents specified in the bidding documents, at the same place as determined in advance in the tender documents. The bid opening shall be presided over by the tender, and all bidders are invited to participate, unseal and read out in public.

18.4.2 Evaluation

Bid committee, set up by the committee according to law, composed of experts in relevant technical and economic fields, is responsible for bid evaluation. They submit a written bid evaluation report to the tenderer and recommend qualified candidates for the bid, or be authorized by the tenderer to directly determine the winning bidder.

18.4.3 Outbid

After the winning bidder is determined, the tender shall issue the notice. After receiving the notice of winning the bid, the successful bidder shall conclude a written contract with the tender in accordance with the tender documents and bidding documents, then performing its obligations in accordance with the contract and complete the winning project.

Words and Phrases

tender	招标
bid	投标,招标
competition	竞争
World Bank	世界银行
international financial institution	国际金融机构
supervision	监督
infrastructure	基础设施
social public interest	社会公共利益
state-owned fund	国有资金
loan	贷款
tender document	招标文件
tender announcement	招标公告
pre-qualification examination	资格预审
business bid	商务标
duration	工期
quotation	报价
award of bid	决标
outbid	中标

Lesson 19 Construction Project Management

19.1 Introduction

Estimating, scheduling and project controls are the major topics in construction project management. The estimate and schedule are developed during the preconstruction phase. Project control systems are set up during the start-up phase and occur primarily in the control phase, but they are documented all the way through the close-out phase. The natural order of progression is shown in Figure 19.1.

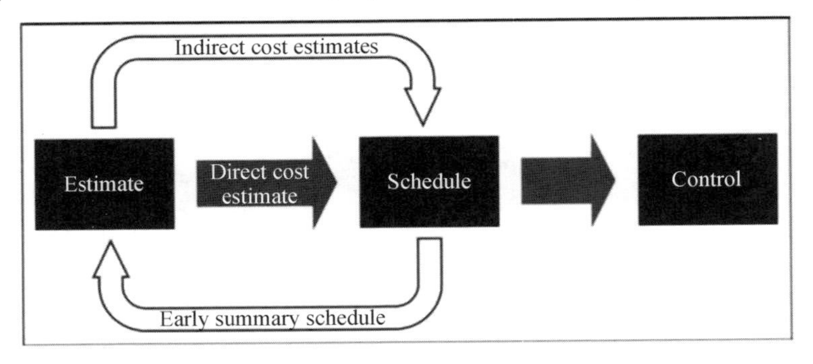

Figure 19.1 The natural order of progression

19.2 Estimating

The three most common types of estimates include: budget estimates, which are performed to identify early budget expectations for the project owner; bid estimates, which include lump sum and heavy-civil unit-price bids are used to participate in competitive bidding; guaranteed maximum price (GMP) proposals, which are popular on many privately negotiated projects.

Cost estimating is the process of collecting, analyzing and summarizing data to prepare an educated projection of the anticipated cost of a project. Costs can be classified into direct costs, indirect costs and markups.

19.2.1 Direct costs

Direct costs directly relate to specific work items. For instance, the cost of building a concrete column—including self-performed labor and materials—would be a direct cost. Similarly, the cost of installing lighting fixtures by an electrical subcontractor—including labor and materials—would be a direct cost.

19.2.2 Indirect costs

Indirect costs refer to either job-site overhead or home-office overhead. Job-site overhead costs are the additional costs of doing construction, which do not directly relate to any specific scope of work but apply to the project as a whole. Examples of jobsite overhead costs include the salaries of the project manager, the superintendent and the project engineers. Home-office overhead costs are the costs for doing business and do not relate to any specific construction project. Examples of home-office overhead costs include the salaries of individuals in the home office and the costs of maintaining the home office.

19.2.3 Markups

Markups are not actual costs. Instead, they refer to the reward of doing business in the form of profit and the risk of adverse circumstances in the form of insurance and contingencies. Contract terms will affect the need for a contractor to incorporate contingencies into their price.

Similarly to the construction of a building, estimating is a logical process consisting of a series of steps. The estimating process applies to all construction firms, including prime and subcontractors. However, the role of prime contractors is broader in nature because they are responsible for acquiring subcontractors and providing a combined price proposal to the project owner.

19.3 Schedule

Many see planning and scheduling as the same operation, but they are slightly different, with proper planning preceding development of the project schedule. Project engineers will not have as much of a role in planning as they will in scheduling but should welcome the opportunity to become involved in development of the project plan whenever it is offered. Planning is usually performed by the project manager and superintendent with the assistance of other experienced and specialized contractor personnel.

19.3.1 Tasks

Planning includes performing the following tasks.

(1) Developing a work breakdown structure (WBS).

(2) Identifying a logic for work activities and restraints on starting or completing activities.

(3) Evaluating the availability of manpower for self-performed work.

(4) Developing a subcontracting plan, including what work is to be subcontracted versus self-performed.

(5) Estimating activity duration from the labor hours on the direct-work pricing recap sheets and from subcontractor input.

(6) Forecasting material and equipment delivery dates.

(7) Identifying owner/architect restraints, such as design package releases, receipt of permits and delivery of owner-supplied equipment.

(8) Selecting means and methods of construction, including choices of concrete formwork (rented versus site-fabricated), internally or externally rented equipment, and hoisting (tower crane or crawler and/or forklift).

19.3.2 Detailed schedule

Many project owners and contractors prefer a detailed schedule to be attached to the contract as an exhibit. Detailed construction schedules are often over 100 line items long, with many stretching to thousands of line items on larger complex projects. The schedule is a construction management tool. The schedule should be detailed enough that progress can be accurately measured, and foremen and subcontractors can develop their three-week look-ahead schedules from it. However, it should not be so detailed that management of the schedule takes on a life of its own.

19.4 Project Controls

Project controls include a set of processes put in place for evaluating the achievement of project success throughout the project delivery. Thus, project controls adopt processes to evaluate the achievement of any of the four stated objectives (safety, cost, time and quality). This approach has progressed from the historical approach of assessing the achievement of cost, schedule and quality by adding a fourth set of processes designed to evaluate safety goals, which is of foremost importance in the construction industry.

19.4.1 Safety control

Most construction firms now predicate the concept of "safety first" in their core business practices. Safety control is accomplished similarly to quality control, with clear expectations and continual open communication. Even if a project has a full-time safety inspector or the corporate safety officer visits the site weekly, most see the general contractor's superintendent as responsible for safety control. General contractors and their superintendents have been found in some situations to be not only contractually responsible for safety, but also legally responsible. If safety violations occur, the superintendent can follow different pathways, which are listed in order of severity: issue a written warning; terminate a direct or subcontractor craftsman; fine or back charge a subcontractor; terminate the subcontractor. The severity of the violation will suggest to the superintendent which pathway is the most appropriate. If accidents occur and someone is injured, there are additional inspections and documentation required.

19.4.2 Cost control

There are five phases of cost control. The first phase begins with an accurate estimate and schedule. All five of the cost control phases are depicted in Figure 19.2, which are estimate (schedule), correct, record, modify and as-built.

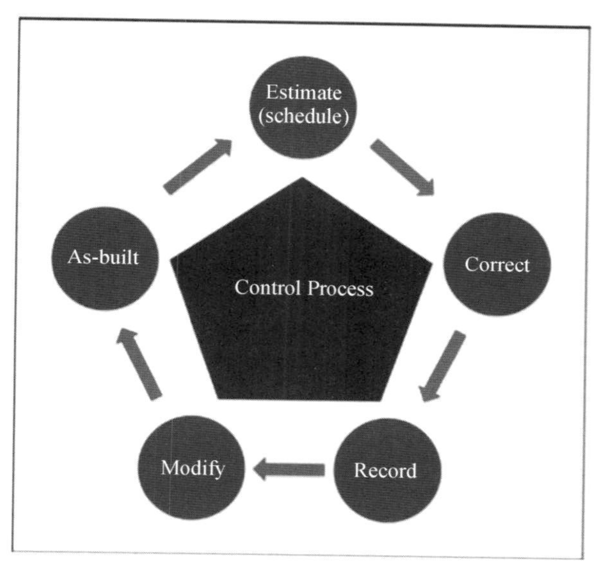

Figure 19.2 Cost/schedule control process

1. Estimate

As introduced earlier, the estimate is an assembly of measured material quantities, competitive market-rate material unit prices, historical direct labor productivity rates, current direct labor wage rates, competitive subcontract quotes, and a series of markups and fees. To better prepare the management team for cost control, the estimate should be assembled by work packages and every line item in the estimate assigned an individual cost code.

2. Correct

No estimate or schedule is perfect. After the contractor receives notice of award of its bid or proposal, the estimate must be corrected with the actual subcontract and purchase order buyout values and entered into the company's cost control system. If the estimator made any mistakes, these must be corrected now, with potential modifications to contingency funds or fees. The superintendent and project engineer cannot begin construction and effectively control costs with an incorrect estimate.

3. Record

This is the largest and most time-consuming phase in the cost control process. The recording of actual costs is also a function most often performed by project engineers. Some of the ways the engineer is involved in recording costs include the followings.

(1) Assisting foremen and superintendents with development of work packages.

(2) Entering direct labor cost codes on time sheets.

(3) Entering cost codes on short- and long-form purchase orders and subcontract agreements.

(4) Verifying that supplier and subcontractor monthly invoice amounts match their contract values.

(5) Entering cost codes on supplier and subcontractor invoices.

(6) Assisting the project manager with monthly fee forecast development.

4. Modify

There is not a perfect estimate nor will many construction projects proceed "exactly" according to the original plan and schedule. Continual comparison of actual costs recorded against the estimate will uncover variances that will warrant attention and potential adjustment by the project team. The contractor cannot wait until the project is finished to find out if it has made money. At that time, there is nothing that can be done to fix the problem. Costs may differ from the estimate for a variety of reasons. Once the reason for the cost overrun is understood, the plan or process may need to be adjusted or modified by the project team.

5. As-built

An as-built estimate, as well as as-built schedule, is an important historical tool that experienced contractors utilize to successfully procure future construction projects. The time to start development of the as-built estimate is when the cost control process is started, with a detailed and cost-coded estimate. During the course of construction, material costs and labor hours are accurately recorded, along with measurement of the actual quantities installed, to develop as-built material unit prices and as-built labor productivity rates. The as-built estimate is defnitely part of the close-out process, but the sooner it is prepared, the more accurate the results will be. These figures should be shared by the project team with staff estimators for incorporation into the contractor's estimating database. In this manner, the last phase of cost control, the as-built phase, is actually the first phase for the next project, and the cost control cycle in Figure 19.2 repeats itself.

19.4.3 Time control

This begins with a detailed and collaboratively developed schedule, not a top-down schedule. Project engineers can be tasked to assist the superintendent with recording actual schedule statistics. Marking up of the contract schedule with actual start and completion dates, along with actual material delivery dates, is integral to development of the as-built schedule. The project superintendent will provide a status report on the schedule at the weekly project team meeting.

Subcontractors will provide the superintendent with their own three-week look-ahead schedules, often at the Monday morning foremen's meeting. A schedule only needs to be revised and reissued if there are major deviations. New schedules must be incorporated into the prime contract and each subcontract and potentially lead to cost ramifications.

19.4.4 Quality control

Before any attempt is made to control quality, an understanding of quality is foremost. The concept of quality is fluid and subjective. This will vary widely, depending on whether the project is for the project owner's headquarters or for a suburban warehouse. In the absence of clear indications, a construction firm would judge quality based on the project conformance, as described in the contractual documents. Therefore, it is important that the project owner conveys their quality expectations through these documents. However, quality control (QC) management first starts with the project owner employing the most qualified design professionals and giving them adequate time and financial resources to prepare a quality design. The next step is for the project owner and designer to clearly communicate their well-defined and reasonable quality expectations in the contract documents. This should all be discussed at the owner–designer–contractor pre-construction meeting. Just as the project owner needs to select a quality GC, the GC needs to select qualified subcontractors and suppliers. Similarly to the design team, all of these firms need to have adequate cost and time allowances, proper materials and equipment, and most importantly, qualified field craftsmen in order to do a quality job.

Words and Phrases

project management	项目管理
estimate	估算
schedule	进度表
preconstruction phase	施工前阶段
control phase	控制阶段
close-out phase	收尾阶段
budget	预算
cost estimate	估价,造价估算,成本估计
direct cost	直接成本,可变成本
markup	加价
labor	劳动力
superintendent	监督员
overhead cost	管理费,非生产费用,杂费,杂项开支
profit	利润
insurance	保险
contingency	偶发事件,意外事故
detailed schedule	详细进度
management tool	管理工具
subcontractor	分包商
contingency fund	应急基金
cost overrun	超支
foremen	工头
quality control	质量控制

foremost	最重要的
conformance	一致
quality expectation	质量检查
field craftsmen	现场工人

How to Write a Scientific Paper

2

The thought of preparing a piece of scientific writing can intimidate even the best writers. However, establishing a suitable mindset and taking an appropriate approach can make the task manageable. Perhaps the most basic is remember that you are writing to communicate, not to impress. Readers of scientific papers want to know what you did, what you found and what it means, they are not seeking great literary merit. Indeed, in scientific writing, readers should notice mainly the content, not the style.

Lesson 20 Norms

Before writing a scientific paper and submitting it to a journal—and indeed, before embarking on your research—you should know the basic ethical norms for scientific conduct and scientific publishing.

20.1 Authenticity and Accuracy

More subtle, and probably more common, are lesser or less definite deviations from accuracy: omitting outlying points from the data reported, preparing figures in ways that accentuate the findings misleadingly, or doing other tweaking. Where to draw the line between editing and distortion may not always be apparent. If in doubt, seek guidance from a more experienced scientist in your field—perhaps one who edits a journal.

The advent of digital imaging has given unethical researchers new ways to falsify findings. (Journal editors, though, have procedures to detect cases in which such falsification of images seems probable.) And ethical researchers may rightly wonder what manipulations of digital images are and are not valid. Sources of guidance in this regard include recent sets of guidelines for use and manipulation of scientific digital images.

20.2 Originality

The findings in a scientific paper must be new. Except in rare and highly specialized circumstances, they cannot have appeared elsewhere in the primary literature. In the few instances in which republication of data may be acceptable—for example, in a more extensive case series or if a paper is republished in another language—the original article must be clearly cited, lest readers erroneously conclude that the old observations are new. To republish a paper (either in another language or for readers in another field), permission normally must be obtained from the journal that originally published the paper.

Originality also means avoiding "salami science"—that is, thinly slicing the findings of a research project, as one might slice a sausage or cuke, in order to publish several papers instead of one (or, in the case of a large research project, many papers instead of a few).

20.3 Credit

If your paper includes information or ideas that are not your own, be sure to cite the source. Likewise, if you use others' wording, remember to place it in quotation marks (or to indent it, if the quoted material is long) and to provide a reference. To avoid inadvertent plagiarism, be sure to include information about the source when you copy or download materials others have written. To avoid the temptation to use others' wording excessively, consider drafting paragraphs without looking directly at the source materials, then look at the materials to check for accuracy.

In journal articles in most fields of science, it is unusual to include quotations from others' work. Rather, authors paraphrase what others have said. Doing so entails truly presenting the ideas in one's own way, changing a word or two does not constitute paraphrasing. On rare occasions—for example, when an author has phrased a concept extraordinarily well—quoting the author's own phrasing may be justified. If you are unsure whether to place in quotation marks a series of words from a publication, do so. If the quotation marks are unnecessary, an editor at the journal can easily remove them. If, however, they are missing but should have been included, the editor might not discover that fact (until, perhaps, a reader later does), or the editor might suspect the fact and send you an inquiry that requires a time-consuming search. Be cautious, and thus save yourself from embarrassment or extra work.

20.4 Ethical Treatment of Humans and Animals

If your research involves human subjects or animals, the journal to which you submit your paper is likely to require documentation that they were treated ethically. Before beginning your study, obtain all needed permissions with regard to human or animal research. Then, in your paper, provide the needed statement(s) in this regard. For guidance, see the instructions to authors for the journal, and use as models papers similar to yours that have appeared in the journal. You may also find it useful to consult relevant sections of style manuals in the sciences. If in doubt, check with the publication office of the journal.

Lesson 21 Writing the Text

21.1 How to Prepare the Title

All words in the title should be chosen with great care, and their association with one another must be carefully managed. Ironically, long titles are often less meaningful than short ones. Without question, most excessively long titles contain "waste" words. Often, these waste words appear right at the start of the title, words such as "Studies on" "Investigations on" and "Observations on". An opening "A" "An" or "The" is also a waste word. Certainly, such words are useless for indexing purposes. Usually, a short version of the title is needed because of space limitations. (The maximum character count is likely to be stated in the journal's instructions to authors.) The title of a paper is a label. It is normally not a sentence, because it is not a sentence, with the usual subject-verb-object arrangement. The meaning and order of the words in the title are important to the potential reader who sees the title in the journal table of contents.

21.2 How to List the Authors and Addresses

The easiest part of preparing a scientific paper is simply entering the bylines: the authors and addresses. If you have co-authors, problems about authorship can range from the trivial to the catastrophic. Commonly, the first author is the person who played the lead role in the research. Qualification to be listed first does not depend on rank. In general, all those listed as authors should have been involved enough to defend the paper or a substantial aspect thereof.

The principles for listing the addresses are simple but often violated. Therefore, authors cannot always be connected with addresses. Most often, however, it has been the style of the journal that creates confusion, rather than sins of commission or omission by the author. The main problem arises when a paper is published by, let us say, three authors from two institutions. In such instances, each author's name and address should include an appropriate designation such as a superior a, b or c after the author's name and before (or after) the appropriate address.

21.3 How to Prepare the Abstract

An abstract should be viewed as a miniature version of the paper. The abstract should provide a brief summary of each of the main sections of the paper: introduction, materials and methods, results, and discussion. An abstract can be defined as a summary of the information in a document. A well-prepared abstract enables readers to identify the basic content of a document quickly and accurately, to determine its relevance to their interests, and thus to decide whether they need to read the document in its entirety. The abstract should state the principal objectives and scope of the investigation, describe the methods employed, summarize the results, state the principal conclusions. The importance of the conclusions is indicated by the fact that they are often given three times: once in the abstract, again in the introduction and again (probably in more detail) in the discussion. Most or all of the abstract should be written in the past tense because it refers to work done.

The preceding rules apply to the abstracts that are used in primary journals and often without change in the secondary services (etc.). This type of abstract is often called an informative abstract, and it is designed to condense the paper. It can and should briefly state the problem, the method used to study the problem, and the principal data and conclusions. Another type of abstract is the indicative abstract (sometimes called a descriptive abstract). This type of abstract is designed to indicate the subjects dealt with in a paper, much like a table of contents, making it easy for potential readers to decide whether to read the paper. However, because of the descriptive rather than substantive nature, it can seldom serve as a substitute for the full paper. When writing the abstract, remember that it will be published by itself, and should be self-contained. That is, it should contain no bibliographic, figure or table references. The language should be familiar to the potential reader.

21.4 How to Write the Introduction

The first section of the text proper should, of course, be the introduction. The purpose of the introduction is to supply sufficient background information to allow the reader to understand and evaluate the results of the present study without needing to refer to previous publications on the topic. The introduction should also provide the rationale for the present study. Choose references carefully to provide the most important background information. Much of the introduction should be written in present tense because you are referring primarily to your problem and the established knowledge relating to it at the start of your work.

Guidelines for a good introduction are as follows.

(1) The introduction should present first, with all possible clarity, the nature and scope of the problem investigated. For example, it should indicate why the overall subject area of the research is important.

(2) It should briefly review the pertinent literature to orient the reader. It also should identify the gap in the literature that the current research was intended to address.

(3) It should then make the objective of the research clear. In some disciplines or journals, it is customary to state here the hypotheses or research questions that the study addressed. In others, the objective may be signaled by wording such as "in order to determine".

(4) It should state the method of the investigation. If deemed necessary, the reasons for the choice of a particular method should be briefly stated.

(5) Finally, in some disciplines and journals, the standard practice is to end the introduction by stating the principal results of the investigation and the principal conclusions suggested by the results. An introduction that is structured in this way has a "funnel" shape, moving from broad and general to narrow and specific. Such an introduction can comfortably funnel readers into reading about the details of your research.

21.5 How to Write the Materials and Methods Section

In "Materials and Methods" (also designated in some cases by other names, such as "Experimental Procedures"), you must give the full details. Most of this section should be written in the past tense. The main purpose of the materials and methods section is to describe (and if necessary, defend) the experimental design and then provide enough detail so that a competent worker can repeat the experiments. Other purposes include providing information that will let readers judge the appropriateness of the experimental methods (and thus the probable validity of the findings) and that will permit assessment of the extent to which the results can be generalized.

21.5.1 Materials

For materials, include the exact technical specifications, quantities and source or method of preparation, sometimes it is even necessary to list pertinent chemical and physical properties of the reagents used. In general, avoid the use of trade names; use of generic or chemical names is usually preferred. This approach avoids the advertising inherent in the trade name. Besides, the nonproprietary name is likely to be known throughout the world, whereas the proprietary name may be known only in the country of origin. However, if there are known differences among proprietary products, and if these differences might be critical, then use of the trade name, plus the name of the manufacturer, is essential. When using trade names, which are usually registered trademarks, capitalize them (for example, Teflon) to distinguish them from generic names. Normally, the generic description should immediately follow the trademark. For example, one would refer to Kleenex facial tissues. In general, it is not necessary to include trademark symbols. However, some journals ask authors to do so. Experimental animals, plants and micro-organisms should be identified accurately, usually by genus, species and strain designations. Sources should be listed and special characteristics described.

21.5.2 Methods

For methods, the usual order of presentation is chronological. Obviously, however, related methods should be described together, and straight chronological order cannot always be followed. For example, even if a particular assay was not done until late in the research, the assay method should be described along with the other assay methods, not by itself in a later part of the materials and methods section.

21.5.3 Headings

The materials and methods section often has subheadings. To see whether subheadings would indeed be suitable—and, if so, what types are likely to be appropriate—look at analogous papers in your target journal. When possible, construct subheadings that "match" those to be used in the results section. The writing of both sections will be easier if you strive for internal consistency, and the reader will be able to grasp quickly the relationship of a particular method to the related results.

21.5.4 Measurements and analysis

Be precise. Methods are similar to cookbook recipes. If a reaction mixture was heated, give the temperature. Questions such as "how" and "how much" should be precisely answered by the author and not left for the reviewer or the reader to puzzle over. Statistical analyses are often necessary, but your paper should emphasize the data, not the statistics. Generally, a lengthy description of statistical methods indicates that the writer has recently acquired this information and believes that the readers need similar enlightenment. Ordinary statistical methods generally should be used without comment; advanced or unusual methods may require a literature citation. In some fields, statistical methods or statistical software customarily is identified at the end of the materials and methods section.

21.5.5 Needs for references

In describing the methods of the investigations, you should give (or direct readers to) sufficient details so that a competent worker could repeat the experiments. If your method is new (unpublished), you must provide all of the needed detail. If, however, the method has been published in a journal, the literature reference should be given. For a method well known to readers, only the literature reference is needed. For a method with which readers might not be familiar, a few words of description tend to be worth adding, especially if the journal in which the method was described might not be readily accessible. If several alternative methods are commonly employed, it is useful to identify your method briefly as well as to cite the reference.

21.5.6 Tables and figures

When many microbial strains or mutants are used in a study, prepare strain tables identifying the source and properties. The properties of multiple chemical compounds can also be presented in tabular form, often to the benefit of both the author and the reader.

Tables can be used for other such types of information. A method, strain, or the like used in only one of several experiments reported in the paper can sometimes be described in the results section. If the description is brief enough, it may be included in a table footnote or figure legend if the journal allows. Figures also can aid in presenting methods. Examples include flow charts of experimental protocols and diagrams of experimental apparatus.

21.5.7 Correct form and grammar

Mistakes in grammar and punctuation are not always serious. The meaning of general concepts, as expressed in the introduction and discussion, can often survive a bit of linguistic mayhem. In materials and methods, however, exact and specific items are being dealt with and precise use of English is a must. Authors are often advised, quite rightly, to minimize use of passive voice. However, in the materials and methods section—as in the current paragraph—passive can often be validly used, for although what was done must be specified, who did it is often irrelevant. Because the materials and methods section usually gives short, discrete bits of information, the writing sometimes becomes telescopic, details essential to the meaning may then be omitted. The most common error is to state the action without, when necessary, stating the agent of the action.

21.6 How to Write the Results

21.6.1 Content of results

There are usually two ingredients of the results section. First, you should give some kinds of overall description of the experiments, providing the big picture without repeating the experimental details previously provided in materials and methods. Second, you should present the data. Your results should be presented in the past tense.

21.6.2 How to handle numbers

If one or only a few determinations are to be presented, they should be treated descriptively in the text. Repetitive determinations should be given in tables or graphs. Any determinations, repetitive or otherwise, should be meaningful. Suppose that in a particular group of experiments, a number of variables were tested (one at a time, of course). Those variables that affect the reaction become determinations or data and, if extensive, are tabulated or graphed. Those variables that do not seem to affect the reaction need not be tabulated or presented. However, it is often important to define even the negative aspects of your experiments. If statistics are used to describe the results, they should be meaningful statistics.

21.6.3 Strive for clarity

The results should be short and sweet, without verbiage. Although the results section is the most important part, it is often the shortest, particularly if it is preceded by a well-written materials and methods section, and followed by a well-written discussion. The results need to be clearly and simply stated because it is the results that constitute the new knowledge that you are contributing to the world. The earlier parts of the paper (introduction, materials and methods) are designed to tell why and how you got the results; the later part of the paper (discussion) is designed to tell what they mean. Obviously, therefore, the whole paper must stand or fall on the basis of the results. Thus, the results must be presented with crystal clarity.

21.7 How to Write the Discussion

The discussion is harder to define than the other sections. Thus, it is usually the hardest section to write. And, whether you know it or not, many papers are rejected by journal editors because of a faulty discussion, even though the data of the paper might be both valid and interesting. Even more likely, the true meaning of the data may be completely obscured by the interpretation presented in the discussion, again resulting in rejection. Many, if not most, discussion sections are too long and verbose, some discussion sections remind one of the diplomat. The main components will be provided if the following injunctions are heeded.

(1) Try to present the principles, relationships and generalizations shown by the results. And bear in mind, in a good discussion, you discuss—do not recapitulate—the results.

(2) Point out any exceptions or any lack of correlation and define unsettled points. Never take the high-risk alternative of trying to cover up or fudge data that do not quite fit.

(3) Show how your results and interpretations agree (or contrast) with previously published work.

(4) Don't be shy. Discuss the theoretical implications of your work, as well as any possible practical applications.

(5) State your conclusions as clearly as possible.

(6) Summarize your evidence for each conclusion. Or, as the wise old scientist will tell you: "Never assume anything except a 4-percent mortgage."

21.8 Ingredient of Acknowledgments

The main text of a scientific paper is usually followed by two additional sections, namely, the acknowledgments and the references.

As to the acknowledgments, two possible ingredients require consideration. First, you should acknowledge any significant technical help that you received from any individual, whether in your laboratory or elsewhere. You should also acknowledge the source of special equipment, cultures, or other materials. Second, it is usually the acknowledgments wherein you should acknowledge any outside financial assistance, such as grants, contracts or fellowships.

21.9 How to Cite the References

There are two rules to follow in the references section, just as in the acknowledgments section. First, list only significant published references. References to unpublished data, abstracts, themes, and other secondary materials should not clutter up the references or literature-cited section. If such a reference seems essential, you may add it parenthetically or, in some journals, as a footnote in the text. A paper that has been accepted for publication can be listed in the literature cited, citing the name of the journal followed by "in press" or "forthcoming". Second, ensure that all parts of every reference are accurate. Doing so may entail checking every reference against the original publication before the manuscript is submitted and perhaps again at the proof stage. Take it from an erstwhile librarian: there are far more mistakes in the references section of a paper than anywhere else.

Don't forget, as a final check, to ensure that all references cited in the text are indeed listed in the literature cited and that all references listed in the literature cited are indeed cited somewhere in the text. Checking that every reference is accurate, and that all cited items appear in the reference list, has become much easier in the electronic era. Common word processing programs include features for tasks such as creating, numbering, and formatting footnotes and endnotes. These features can aid in citing references and developing reference lists. Some journals, however, say not to use these features, which can interfere with their publishing process. Check the journal's instructions to authors in this regard.

Lesson 22 Preparing the Tables and Figures

22.1 When to Use Tables

As a rule, do not construct a table unless repetitive data must be presented. There are two reasons for this general rule. First, it is simply not good science to regurgitate reams of data just because you have them in your laboratory notebooks, only samples and breakpoints need to be given. Second, the cost of publishing tables can be high compared with that of text, and all of us involved with the generation and publication of scientific literature should worry about the cost.

If you made (or need to present) only a few determinations, give the data in the text. Table 22.1 is useless, yet they are typical of many tables that are submitted to journals. Table 22.1 is faulty because two of the columns give standard conditions, not variables and not data. If temperature is a variable in the experiments, it can have its column. The data presented in the table can be given in the text itself in a form that is readily comprehensible to the reader, without taking up space with a table.

Table 22.1 Effect of aeration on growth of streptomyces coelicolor

Temperature /℃	Number of experiment	Aeration of growth medium	Growth
24	5	+	78
24	5	-	0

22.2 How to Arrange Tabular Material

Since a table has both left-right and up-down dimensions, you have two choices. The data can be presented either horizontally or vertically. But "can" does not mean "should". The data should be organized so that the like elements read down, not across.

The point about ease for the reader would seem to be obvious. The point about reduced printing costs refers to the fact that all columns must be wide or deep in the across arrangement because of the diversity of elements, whereas some columns numbers can be narrow without runovers in the down arrangement. Thus, Table 22.2 appears to be smaller than Table 22.3, although it contains the same information.

Table 22.2 Characteristics of antibiotic-producing streptomyces(1)

Determination	Streptomyces fluoricolor	Streptomyces griseus	Streptomyces coelicolor	Streptomyces nocolor
Optimal growth temperature/℃	-10	24	28	92
Color of mycelium	Tan	Gray	Red	Purple
Antibiotic-produced	Fluricillinmycin	Streptomycin	Rholmondelay	Nomycin

Table22.2 (continued)

Determination	Streptomyces fluoricolor	Streptomyces griseus	Streptomyces coelicolor	Streptomyces nocolor
Yield of antibiotic /(mg · mL^{-1})	4,108	78	2	0

Table 22.3 Characteristics of antibiotic-producing streptomyces (2)

Organism	Optimal growth temperature/℃	Color of mycelium	Antibiotic-produced	Yield of antibiotic /(mg · mL^{-1})
Streptomyces fluoricolor	-10	Tan	Fluricillinmycin	4,108
Streptomyces griseus	24	Gray	Streptomycin	78
Streptomyces coelicolor	28	Red	Rholmondelay	2
Streptomyces nocolor	92	Purple	Nomycin	0

22.3 Exponents in Table Headings

If possible, avoid using exponents in table headings. Confusion has resulted because some journals use positive exponents and some use negative exponents to mean the same thing. For example, some have used "cpm × 10^3" to refer to thousands of counts per minute, whereas others have used "cpm× 10^{-3}" for the same thousands of counts. If it is not possible to avoid such labels in table headings (or in figures), it may be worthwhile to state in a footnote (or in the figure legend), in words that eliminate the ambiguity, what convention is being used.

22.4 Following Journal's Instructions

Instructions to authors commonly include a section about tables. Before preparing your tables, check the instructions to authors of your target journal. These instructions may indicate such items as the dimensions of the space available, the symbols or form of lettering for indicating footnotes to tables, and the electronic tools to use in preparing tables. Looking at tables in the journal as examples also can aid in preparing suitable tables. Style manuals in the sciences provide guidance in preparing not only text but also tables and figures. If your target journal specifies a style manual that it follows, consult it in this regard. Even if the journal does not specify a style manual, looking at one relevant to your field can aid in preparing effective tables and figures.

22.5 Titles, Footnotes and Abbreviations

The title of the table (or the legend of a figure) is like the title of the paper itself. That is, the title or legend should be concise and not divided into two or more clauses or sentences. Unnecessary words should be omitted. Give careful thought to the footnotes to your tables. If abbreviations must be defined, you can often give all or most of the definitions in the first table. Then later tables can carry the simple footnote: "Abbreviations as in Table 1". To identify abbreviations that your target journal considers acceptable in tables, you can look at tables published in the journal. Also, some journals list in their instructions to authors the abbreviations that can be used without definition in tables that they publish.

Lesson 23 How to Prepare Effective Graphs

23.1 When to Use Graphs

Graphs resemble tables as a means of presenting data in an organized way. In fact, the results of many experiments can be presented either as tables or as graphs. This is often a difficult decision. A good rule might be this: if the data show pronounced trends, making an interesting picture, use a graph. If the numbers just sit there, with no exciting trend in evidence, a table should be satisfactory. Tables are also preferred for presenting exact numbers. Examine Table 23.1 and Figure 23.1, both of which record exactly the same data. Either format would be acceptable for publication, but Figure 23.1 clearly seems superior to Table 23.1. In the figure, the synergistic action of the two drug combinations is immediately apparent. Thus, the reader can quickly grasp the significance of the data. It also appears from the graph that streptomycin is more effective than is isoniazid, although its action is somewhat slower, this aspect of the results is not readily apparent from the table.

Table 23.1 Effects on mycobacterium tuberculosis

Treatment	Percentage of negative cultures at			
	2 weeks	4 weeks	6 weeks	8 weeks
Streptomycin	5	10	15	20
Isoniazid	8	12	15	15
Streptomycin+Isoniazid	30	60	80	100

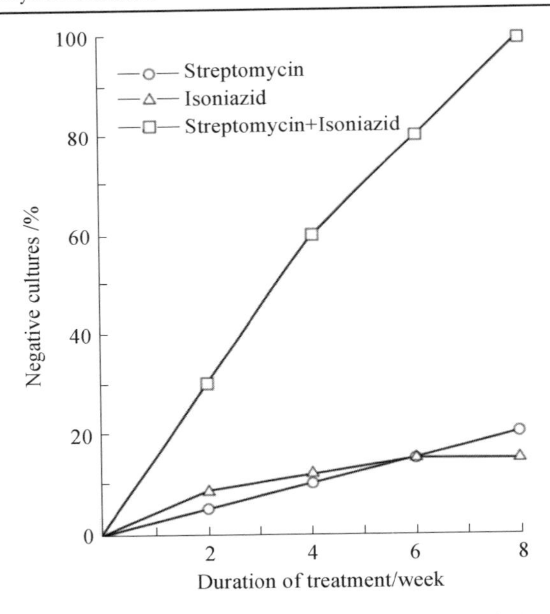

Figure 23.1 Effects on mycobacterium tuberculosis

23.2 How to Prepare Graphs

Today, we prepare graphs by computer. However, the principles of producing good graphs have not changed. The sizes of the letters and symbols, for example, must be chosen so that the final published graph in the journal is clear and readable. The size of the lettering must be based on the anticipated reduction that will occur in the publishing process. This factor can be especially important if you are combining two or more graphs into a single illustration. Remember: text that is easy to read on a large computer screen may become illegible when reduced to the width of a journal column.

If your paper contains two or more graphs that are most meaningful when viewed together, consider grouping them in a single illustration. To maximize readability, place the graphs above and below each other rather than side by side. For example, in a 2-column journal, placing three graphs in an "above and below" arrangement allows each graph to be one or two columns in width. If the graphs appear side by side, each can average only 1/3 of a page wide.

Whether or not you group graphs in such a composite arrangement, be consistent from graph to graph. For example, if you are comparing interventions, keep using the same symbol for the same intervention. Also be consistent in other aspects of design. Both conceptually and aesthetically, the graphs in your paper should function as a set.

23.3 Symbols and Legends

If there is a space in the graph itself, use it to present the key to the symbols. If you must define the symbols in the figure legend, you should use only those symbols that are considered standard and that are widely available. Perhaps the most standard symbols are open and closed circles, triangles and squares. If you have just one curve, use open circles for the reference points, use open triangles for the second, open squares for the third, closed circles for the fourth, etc. If you need more symbols, you probably have too many curves for one graph, and you should consider dividing it into two. Different types of connecting lines (solid, dashed) can also be used. But do not use different types of connecting lines and different symbols.

As to the legends, they should normally be provided on a separate page, not at the bottom or top of the illustrations themselves. The main reason is that the two portions are commonly processed separately during journal production. Consult the instructions to authors of your target journal regarding this matter and other requirements for graphs.

23.4 A Few Tips on Graphs

Note that some journals (mainly the larger and wealthier ones) redraw graphs and some other types of figures to suit their own format. Whether or not a journal will do so, prepare your graphs well. Doing so will help make your findings and their value clear and will help show the care with which you do your work.

Lesson 24 How to Prepare Effective Photographs

If your paper is to be illustrated with one or more photographs, there are several factors to keep in mind. As with graphs, the size (especially width) of the photograph in relation to the column and page width of the journal can be important. Try to avoid dimensions that will require excessive reduction of the photo graph to suit the journal page.

24.1 Submission Formats

Today, journals normally request photo graphs in electronic format. To ascertain requirements for photographs, see the instructions to authors for your target journal. For example, check what formats (such as EPS, JPEG or TIFF) are acceptable and what resolution is required.

24.2 Cropping

Whatever the quality of your photographs, you want to have them published legibly. To some degree, you can control this process yourself if you use your head. If you are concerned that detail might be lost by excessive reduction, there are several ways you might avoid this. Seldom do you need the whole photograph, right out to all four edges. Therefore, crop the photograph to include only the important part. Commonly, photographs are cropped digitally. If you are submitting a print, you can write "crop marks" on the margin to show where the photograph should be cropped.

24.3 Necessary Keys and Guides

If you can't crop down to the features of special interest, consider superimposing arrows or letters on the photographs. In this way, you can draw the reader's attention to the significant features. Having arrows or letters to refer to can aid in writing clear, concise legends. Unless your journal requests that photographs and other illustrations be embedded in the text, it is a good idea to indicate the preferred location for each illustration. In this way, you will be sure that all illustrations have been referred to in the text, in one-two-three order, and the printer will know how to weave the illustrations into the text so that each one is close to the text related to it.

With electron micrographs, put a micrometer marker directly on the micrograph. In this way, regardless of any reduction (or even enlargement) in the printing process, the magnification factor is clearly evident. The practice of putting the magnification in the legend is not advisable, and some journals no longer allow it, precisely because the size is likely to change in printing. In other photographs where the size of the object is important, likewise include a scale bar. Sometimes showing a familiar object, such as a paper clip, near the object can help readers discern an object's size.

24.4 Color

More color illustrations are appearing, and use of color has become relatively common in some fields and journals. If you have the option of including color, consider whether doing so will improve your scientific paper. If you are considering using color, see the instructions to authors of your target journal for specifications regarding color illustrations and for information on any charges for color. If color illustrations are to be printed, authors commonly must pay a fee. Some journals, however, do not charge for color.

24.5 Line Drawing

In some fields, line drawings are superior to photographs in showing important details. Such illustrations are also common in medicine, especially in presenting anatomic views, and indeed have become virtually an art form. When illustrations are necessary, the services of a professional illustrator generally are required. Such illustrators are available at many universities and other research institutions and can be identified through associations of scientific and medical illustrators.

3 Common English Terms in Civil Engineering

A

acceleration	加速度
acceptable quality	合格质量
acceptance	验收
acceptance lot	验收批量
accidental action	偶然作用
accidental combination	偶然组合
accidental combination for action effects	作用效应偶然组合
accidental load	偶然荷载
accidental situation	偶然状况
action	作用
admixture	外加剂
aggregate	骨料

air content	含气量
air-dried timber	气干材
allowable slenderness ratio of steel member	钢构件容许长细比
allowable slenderness ratio of timber compression member	受压木构件容许长细比
allowable stress range of fatigue	疲劳容许应力幅
allowable subsoil deformation	地基变形允许值
allowable ultimate tensile strain of reinforcement	钢筋拉应变限值
allowable value of crack width	裂缝宽度容许值
allowable value of deflection of structural member	构件挠度容许值
allowable value of deflection of timber bending member	受弯木构件挠度容许值
allowable value of deformation of steel member	钢构件变形容许值
allowable value of deformation of structural member	构件变形容许值
ambient temperature	环境温度
amplified coefficient of eccentricity	偏心距增大系数
amplitude of vibration	振幅
anchorage	锚具
anchorage length of steel bar	钢筋锚固长度
approval analysis during construction stage	施工阶段验算

arch	拱
arch-shaped roof truss	拱形屋架
area of shear plane	剪面面积
area of transformed section	换算截面面积
arch structure	拱结构
area of section	截面面积
seismic joint	防震缝
seismic design	建筑抗震设计
assembled monolithic concrete structure	装配整体式混凝土结构
assembly part	部件
a separate waterproof barrier	一道防水设防
at an angle to grain	斜纹
automatic welding	自动焊接
auxiliary steel bar	架立钢筋

B

batten plate	缀板
balanced depth of compression zone	界限受压区高度
balanced eccentricity	界限偏心距
bar splice	钢筋接头
bark pocket	夹皮
base course	基层

basic variable	基本变量
beam fixed at both ends	两端固定梁
beam	梁
bearing capacity	承载能力
bearing plane of notch	齿承压面
bearing plate	支承板
bearing stiffener	支承加劲肋
bending moment	弯矩
bent frame	排架
bent-up steel bar	弯起钢筋
block	砌块
block masonry	砌块砌体
block masonry structure	砌块砌体结构
blow hole	气孔
board	板材
bolt	螺栓
bolted connection	(钢结构)螺栓连接
bolted joint	(木结构)螺栓连接
bolted steel structure	螺栓连接钢结构
bonded prestressed concrete structure	有黏结预应力混凝土结构
bow	顺弯
box foundation	箱形基础
bracing member	支管
brake member	制动构件
breadth of section	截面宽度
breadth of wall between windows	窗间墙宽度

brick masonry	砖砌体
brick masonry column	砖砌体柱
brick masonry structure	砖砌体结构
brick masonry wall	砖砌体墙
brittle failure	脆性破坏
brittle fracture	脆断
broad-leaved wood	阔叶树材
buckling	屈服
building	房屋建筑
building and civil engineering structure	建筑及土木工程结构
building decoration	建筑装饰装修
building engineering	建筑施工
building engineering	房屋建筑工程
building ground	建筑地面
building height	房屋高度
building structural material	建筑结构材料
building structural unit	建筑结构单元
building structure	建筑结构
built-up member	组合构件
built-up steel column	格构式钢柱
bundled tube structure	成束筒结构
burn-through	烧穿
butt connection; butt joint	对接
butt weld	对接焊缝

C

cable-suspended structure	悬索结构
caisson foundation	沉箱基础
calculating area of compression member	受压构件计算面积
calculation of load-carrying capacity of member	构件承载能力计算
calculating overturning point	计算倾覆点
calibration	校准法
camber of structural member	结构构件起拱
cantilever beam; cantilever beam	挑梁;悬臂梁
cap of reinforced concrete column	钢筋混凝土柱帽
carbonation of concrete	混凝土碳化
cast-in-situ concrete slab column structure	现浇板柱结构
cast-in-situ concrete structure	现浇混凝土结构
category of construction quality control	施工质量控制等级
cavity wall filled with insulation	夹心墙
cement	水泥
cement content	水泥含量
cement deep mixing	水泥土搅拌法

cement mortar	水泥砂浆
characteristic combination	标准组合
characteristic value/nominal value	标准值
characteristic/nominal combination	标准组合
characteristic value of a property of a material	材料性能标准值
characteristic value of an action	作用标准值
characteristic value of concrete compressive strength	混凝土轴心抗压强度标准值
characteristic value of concrete tensile strength	混凝土轴心抗拉标准值
characteristic value of cubic concrete compressive strength	混凝土立方体抗压强度标准值
characteristic value of earthquake action	地震作用标准值
characteristic value of horizontal crane load	吊车水平荷载标准值
characteristic value of live load on floor or roof	楼面、屋面活荷载标准值
characteristic value of masonry strength	砌体强度标准值
characteristic value of material strength	材料强度标准值
characteristic value of permanent action	永久作用标准值
chimney	烟囱

chord member	主管
circular double-layer suspended cable	圆形双层悬索
circular single-layer suspended cable	圆形单层悬索
circumferential weld	环形焊缝
civil architecture；civil building	民用建筑
civil engineering	土木工程
clincher	扒钉
coefficient for combination value of actions	作用组合值系数
coefficient of effects of actions	作用效应系数
coefficient of friction	摩擦系数
cold bend inspection of steel bar	冷弯试验
column bracing	柱间支撑
cold drawn bar	冷拉钢筋
cold drawn wire	冷拉钢丝
cold-formed thin-walled section steel	冷弯薄壁型钢
cold-formed thin-walled steel structure	冷弯薄壁型钢结构
cold-rolled deformed bar	冷轧带肋钢筋
column	柱
combination for action effects	作用效应组合
combination for long-term action effects	长期效应组合

combination for short-term action effects	短期效应组合
combination value	组合值
combination value of actions	作用组合值
combined course	结合层
combined footing	联合基础
common defect	一般缺陷
compaction	密实度
compliance control	合格控制
component	部件
composite brick masonry member	组合砖砌体构件
compositefloor system	组合楼盖
compositefloor with profiled steel sheet	压型钢板楼板
compositemortar	混合砂浆
compositeroof truss	组合屋架
compositemember	组合构件
composite rubber and steel support	橡胶支座
composite steel and concrete beam	钢与混凝土组合梁
compressive capacity	受压承载能力
compression member with large eccentricity	大偏心受压构件
compression member with small eccentricity	小偏心受压构件
compressive strength	抗压强度
compound stirrup	复合箍筋

concentration of plastic deformation	塑性变形集中
conceptual earthquake-resistant design	建筑抗震概念设计
concrete	混凝土
concrete column	混凝土柱
concrete consistence	混凝土稠度
concrete-filled steel tubular member	钢管混凝土构件
concrete folded-plate structure	混凝土折板结构
concrete foundation	混凝土基础
concrete mix ratio	混凝土配合比
concrete small hollow block	混凝土小型空心砌块
concrete structure	混凝土结构
concrete wall	混凝土墙
conifer; coniferous wood	针叶树材
connecting plate	连接板
connection	连接
construction joint	施工缝
connections of steel structure	钢结构连接
connections of timber structure	木结构连接
consistency of mortar	砂浆稠度
constant cross-section column	等截面柱
construction work	建筑物(构筑物)
continuous beam	连续梁
continuous seam	通缝

continuous weld	连续焊缝
control grade of construction quality	施工质量控制等级
control point	控制缝
core area of section	截面核心面积
core column	芯柱
core tube supported structure	核心筒悬挂结构
corrosion of steel bar	钢筋锈蚀
counting inspection	计数检验
coupled wall	连肢墙
coupling wall-beam	连梁
coupling wall-column	墙肢
coursing degree of mortar	砂浆分层度
cover plate	盖板
covered electrode	焊条
crack	裂缝
crack resistance	抗裂度
crack width	裂缝宽度
crane girder	吊车梁
crane load	吊车荷载
creep of concrete	混凝土徐变
crook	横弯
cross beam	井字梁
crosswise shrinkage crack	横向缩缝
cup	翘弯
curved support	弧形支座

cushion	换填垫层法
cyclic action	多次重复作用
cylindrical brick arch	砖筒拱
cylinder pile foundation; cylinder caisson foundation	管柱基础

D

damp	阻尼
decay	腐朽
decay prevention of timber structure	木结构防腐
deep beam	深梁
deep flexural structure	深受弯构件
deep mixing	深层搅拌法
defect	缺陷
defect in timber	木材缺陷
deflection	挠度
deformation	变形
deformation analysis	变形验算
deformation modulus	变形模量
degree of freedom	自由度
degree of gravity vertical for structure or structural member	结构构件垂直度

degree of gravity vertical for wall surface	墙面垂直度
degree of plainness for structural member	构件平整度
degree of plainness for wall surface	墙面平整度
degree of reliability	可靠度
depth of neutral axis	中和轴高度
depth of notch	齿深
depth of section	截面高度
design basic acceleration of ground motion	设计基本地震加速度
design characteristic period of ground motion	设计特征周期
designations of steel	钢材牌号
design of building and civil engineering structure	工程结构设计
design of building structure	建筑结构设计
design parameters of ground motion	设计地震动参数
design reference period	设计基准期
design situation	设计状况
design value of a load	荷载设计值
design value of an action	作用设计值
design value of a property of a material	材料性能设计值

design value of earthquake-resistant strength of materials	材料抗震强度设计值
design value of load-carrying capacity of members	构件承载能力设计值
design value of material strength	材料强度设计值
design value of strength	强度设计值
design working life	设计使用年限
destructive test	破损试验
detail	细部
detailing reinforcement	构造配筋
detailing requirement	构造要求
details of seismic design	抗震构造措施
deterministic method	定值设计法
diameter of section	截面直径
dimension lumber	规格材(木材)
diaphragm	横隔板
dimensional error	尺寸偏差
distribution factor of snow pressure	屋面积雪分布系数
displacement	位移
dividing joint	分格法
dominant item	主控项目
double component concrete column	双肢柱
down-stayed composite beam	下撑式组合梁
dry jet mixing	粉体喷搅法
ductile failure	延性破坏
ductile frame	延性框架

durability	耐久性
dynamic action	动态作用
dynamic coefficient	动力系数
dynamic compaction; dynamic consolidation	强夯法
dynamic consolidation foundation	强夯地基
dynamic design	动态设计
dynamic effect factor	动态作用系数
dynamic moment of inertia	转动惯量
dynamic replacement	强夯置换法

E

earth pile	土挤密桩法
earth pressure	土压力
earthquake	地震
earthquake action	地震作用
earthquake epicentre	震中
earthquake focus	震源
earthquake intensity	地震烈度
earthquake magnitude	地震震级
earthquake-resistant design	抗震设计
earthquake-resistant detailing requirement	抗震构造要求

earthquake zone	地震区
eccentricity	偏心矩
effective area of fillet weld	角焊缝有效面积
effective cross-section area of bolt	螺栓有效截面面积
effective depth of section	截面有效高度
effective height	计算高度
effective length	计算长度
effective length of fillet weld	角焊缝有效计算长度
effective length of nail	钉有效长度
effective span	计算跨度
effective width	有效宽度
effective width factor	有效宽度系数
effect of action	作用效应
elastic analysis scheme	弹性方案
elastic deformation	弹性变形
elastic foundation beam	弹性地基梁
elastic foundation plate	弹性地基板
elastically supported continuous girder	弹性支座连续梁
elasticity modulus	弹性模量
elasticity modulus of material	材料弹性模量
element	构件
elevated overhead roof	架空屋面
elongation rate	伸长率
embedded part	预埋件
entrapped air	含气量

equilibrium moisture content	平衡含水率
equivalent slenderness ratio	换算长细比
equivalent uniformly distributed live load	等效均布活荷载
Euler's critical load	欧拉临界力
Euler's critical stress	欧拉临界应力
excessive penetration	塌陷
expansion and contraction joint	伸缩缝
explosion action	爆炸作用

F

fatigue capacity	疲劳承载能力
fatigue strength	疲劳强度
fiber concrete	纤维混凝土
filler course	填充层
filler plate	填板门
fillet weld	角焊缝
final setting time	终凝时间
finger joint	指接
fired common brick	烧结普通砖
fired perforated brick	烧结多孔砖
first moment of area	截面面积矩
first order elastic analysis	一阶弹性分析

fish eye	白点
fish-belly beam	角腹式梁
fixed action	固定作用
flange plate	翼缘板
flexible connection	柔性连接
flexible waterproof layer	柔性防水层
floor live load；roof live load	楼面、屋面活荷载
floor plate	楼板
floor system	楼盖
four sides（edges）supported plate	四边支承板
folded-plate structure	折板结构
force per unit area	面分布力
force per unit length	线分布力
force per unit volume	体分布力
force（weight）density	重力密度
foundation	基础
foundation earth layer	基土
frame	框架
frame braced with strong bracing system	强支撑框架
frame braced with weak bracing system	弱支撑框架
frame structure	框架结构
frame tube structure	框架-筒体结构
free action	自由（可动）作用
frequency	频率

frequent combinations	频遇组合
frequent value	频遇值
friction coefficient of masonry	砌体摩擦系数
full degree of mortar at bed joint	砂浆饱满度
function of acceptance	验收函数
fundamental combination	基本组合
fundamental combination for action effects	作用效应基本组合

G

gang nail plate joint	钉板连接
gap joint	间隙节点
general item	一般项目
girder	梁
glue used for structural timber	木结构用胶
glued joint	胶合接头
glued laminated timber(glulam)	层板胶合木
glued laminated timber structure	层板胶合结构
glued lumber	胶合材
grade of waterproof	防水等级
gravity density	重力密度(重度)
grip	夹具
groove	坡口

ground treatment	地基处理
grout for concrete small hollow	混凝土砌块灌孔
grouting foundation	注浆地基
gusset plate	节点板

H

handing over inspection	交接检验
hanger	吊环
hanging steel bar	吊筋
heat fusion method	热熔法
heat tempering bar	热处理钢筋
heartwood	芯材
heavy tamping foundation	重锤夯实地基
height of section	截面高度
height variation factor of wind pressure	风压高度变化系数
high-rise structure	高耸结构
high-strength bolt	高强度螺栓
high-strength bolted bearing type join	承压型高强度螺栓连接
high-strength bolted connection	高强度螺栓连接
high-strength bolted friction-type joint	摩擦型高强度螺栓连接

hinge support	铰轴支座
hinged connection	铰接
hollow brick	空心砖
hollow ratio of masonry unit	块体空心率
honeycomb	蜂窝
hook	弯钩
hoop	箍筋
hot air welding method	热风焊接法
hot-rolled deformed bar	热轧带肋钢筋
hot-rolled plain bar	热轧光圆钢筋
hot-rolled section steel	热轧型钢

I

impact toughness	冲击韧性
impermeability	抗渗性
imposed deformation	外加变形
impounded roof	蓄水屋面
inclined section	斜截面
inclined stirrup	斜向箍筋
incomplete penetration	未焊透
incompletely filled groove	未焊满
indented wire	刻痕钢丝
industrial building	工业建筑

initial control	初步控制
insect prevention of timber structure	木结构防虫
inspection	检验
inspection at original space	原位检测
inspection for properties of glue used in structural member	结构用胶性能检验
inspection for properties of mortar	砂浆性能检验
inspection for properties of steel bar	钢筋性能检验
inspection lot	检验批
inspection of structural performance	结构性能检验
intermediate assembled structure	中拼单元
intermediate stiffener	中间加劲肋
intermittent weld	断续焊缝
inversion type roof	倒置式屋面

J

isolating course	隔离层
jet grouting	高压喷射注浆法
jet grouting foundation	高压喷射注浆地基
joint	节点

joint of reinforcement 钢筋接头

K

key joint 键连接
kinetic design 动态设计
knot 节子(木节)

L

laced of battened compression member 格构式钢柱
lacing and batten element 缀材(缀件)
lacing bar 缀条
laminated strand lumber(LSL) 层叠木片胶合木
laminated veneer lumber(LVL) 旋切板胶合木
lamination 层板
lap connection 叠接(搭接)
lapped length of steel bar 钢筋搭接长度
large panel concrete structure 混凝土大板结构
large-form concrete structure 大模板结构
lateral bending 侧向弯曲

lateral displacement stiffness of story	楼层侧移刚度
lateral displacement stiffness of structure	结构侧移刚度
lateral force resistant wall structure	抗侧力墙体结构
leaning column	摇摆柱
leg size of fillet weld	角焊缝焊脚尺寸
length	长度
length of shear plane	剪面长度
lengthwise shrinkage crack	纵向缩缝
life of water proof layer	防水层合理使用年限
lift-slab structure	升板结构
light weight aggregate concrete	轻骨料混凝土
light wood frame construction	轻型木结构
lime pile	石灰桩法
lime soil pile	灰土挤密桩法
limit of acceptance	验收界限
limit state equation	极限状态方程
limit state	极限状态
limit state method	极限状态设计法
limiting design value	设计限值
limiting value for local dimension of masonry structure	砌体结构局部尺寸限值
limiting value for sectional dimension	截面尺寸限值
limiting value for supporting length	支承长度限值

limiting value for total height of masonry structure	砌体结构总高度限值
linear strain	线应变
linear expansion coefficient	线膨胀系数
lintel	过梁
liquefaction of saturated soil	砂土液化
load	荷载
load bearing wall	承重墙
load combination	荷载组合
load effect	荷载效应
load effect combination	荷载效应组合
load-carrying capacity	承载能力
log	原木
log timber structure	原木结构
long term rigidity of member	构件长期刚度
longitude horizontal bracing	纵向水平支撑
longitudinal steel bar	纵向钢筋
longitudinal stiffener	纵向加劲肋
longitudinal weld	纵向焊缝
losses of prestress	预应力损失
low frequency cyclic action	低周反复作用
lump material	块体

M

machine stress-graded lumber	木材机械分级
main axis	强轴
main beam	主梁
manual welding	手工焊接
manufacture control	生产控制
map cracking	龟裂
masonry	砌体
masonry-concrete structure	砖混结构
masonry lintel	砖过梁
masonry member	无筋砌体构件
masonry structure	砌体结构
masonry-timber structure	砖木结构
masonry units	块体
mass density	质量密度
mechanical properties of materials	材料力学性能
melt-thru	烧穿
member	构件
method of sampling	抽样方法
minimum strength class of masonry	砌体材料最低强度等级
mix ratio of mortar	砂浆配合比
mixed structure	混合结构

mixing water	拌合水
mode of vibration	振型
moment of area; moment of inertia	截面惯性矩
elasticity modulus	弹性模量
moment of momentum	动量矩
monitor frame	天窗架
mortar	砂浆
mortar for concrete small hollow	混凝土砌块砌筑砂浆
multi-defense system of earthquake-resistant building	多道设防抗震建筑
multi-tube supported suspended structure	多筒悬挂结构

N

nailed joint	钉连接
natural frequency	自振(固有)频率
natural period of vibration	自振周期
net height	净高
net span	净跨度
net water/cement ratio	净水灰比
nodal bracing force	支撑力
non-destructive inspection of weld	焊缝无损检验

non-destructive test	非破损检验
non-load-bearing wall	非承重墙
non-reinforced spread foundation	无筋扩展地基
non-uniform cross-section beam	变截面梁
non-uniformly distributed strain coefficient of longitudinal tensile reinforcement	纵向受拉钢筋应变不均匀系数
normal concrete	普通混凝土
normal force	轴向力
normal section	正截面
normal stress	正应力
normal value of geometric parameter	几何参数标准值
normalized web slenderness	通用高厚比
number of sampling	抽样数量

O

open caisson foundation	沉井基础
open web roof truss	空腹屋架
ordinary concrete	普通混凝土
ordinary steel bar	普通钢筋
outstanding width of flange	翼缘板外伸宽度

outstanding width of stiffener	加劲肋外伸宽度
overall stability	整体稳定
over-all stability reduction coefficient of steel beam	钢梁整体稳定系数
overlap	焊瘤
overlap joint	搭接节点
overturning or slip resistance analysis	抗倾覆、滑移验算

P

padding plate	垫板
panel zone of column web	柱腹板节点域
parallel strand lumber(PSL)	平行木片胶合木
parallel to grain	顺纹
part	零件
partial penetrated butt weld	不焊透对接焊缝
partial safety factor	分项系数
partial safety factor for action	作用分项系数
partial safety factor for property of a material	材料性能分项系数
partial safety factor for resistance	抗力分项系数
partition	非承重墙
penetrated butt weld	透焊对接焊缝

penetration	贯入度
percentage of reinforcement	配筋率
perforated brick	多孔砖
perimeter of section	截面周长
period	周期
periodic vibration	周期振动
permanent action	永久作用
permanent load	永久荷载
permissible (allowable) stresses method	容许应力设计法
perpendicular to grain	横纹
persistent situation	持久状况
pile	桩
pile foundation	桩基础
plain concrete structure	素混凝土结构
plane hypothesis	平截面假定
plane structure	平面结构
plane trussed lattice grid	平面桁架系网架
plank	板材
planted roof	种植屋面
plastic deformation	塑性变形
plastic hinge	塑性铰
plate	板
plate-like space frame	干板型网架
plate-like space truss	平板型网架
plug weld	塞焊缝

plywood	胶合板
plywood structure	胶合板结构
pneumatic structure	充气结构
pockmark	麻面
Poisson ratio	泊松比
polar second moment of area; polar moment of inertia	截面极惯性矩
polygonal top-chord roof truss	多边形屋架
post-buckling strength of web plate	腹板屈服后强度
post-tensional prestressed concrete structure	后张法预应力混凝土结构
precast reinforced concrete member	预制混凝土构件
prefabricated concrete structure	装配式混凝土结构
preloading	预压法
preloading foundation	预压地基
pressed pile by anchor rod	锚杆静压桩
presetting time	初凝时间
prestress	预应力
prestressed concrete structure	预应力混凝土结构
prestressed steel structure	预应力钢结构
pre-tensional prestressed concrete structure	先张法预应力混凝土结构
primary bracing	初期支护
primary control	初步控制
primary structure	基体

principal strain	主应变
principal stress	主应力
probabilistic method	概率设计法
properties of fresh concrete	可塑混凝土性能
properties of hardened concrete	硬化混凝土性能
property of building structural materials	建筑结构材料性能

Q

quality grade of structural timber	木材质量等级
quality grade of weld	焊缝质量级别
quality inspection of bolted connection	螺栓连接质量检验
quality inspection of masonry	砌体质量检验
quality inspection of riveted connection	铆钉连接质量检验
quasi-permanent value of live load on floor or roof	楼面、屋面活荷载准永久值
quality of appearance	观感质量
quality of building engineering	建筑工程质量
quantitative inspection	计量检验
quasi-permanent value of an action	作用准永久值

R

radial check	辐裂
radius of gyration	截面回转半径
raft foundation	筏形基础
rammed soil-cement pile	夯实水泥土桩法
ratio of reinforcement	配筋率
ratio of shear span to effective depth of section	剪跨比
redistribution of internal force	内力重分布
reducing coefficient of live load	活荷载折减系数
reference snow pressure	基本雪压
reference wind pressure	基本风压
regular earthquake-resistant building	规则抗震建筑
reinforced concrete masonry shear wall structure	配筋砌块砌体剪力墙结构
reinforced concrete deep beam	混凝土深梁
reinforcement ratio	配筋率
reinforcement ratio per unit volume	体积配筋率
relative eccentricity	偏心率
relaxation of prestressed tendon	预应筋松弛

reliability	可靠性
reliability index	可靠指标
repair	返修
repeated action	多次重复作用
representative value of an action	作用代表值
representative values of a load	荷载代表值
representative value of gravity load	重力荷载代表值
resistance	抗力
resistance to abrasion	耐磨性
resistance to water penetration	抗渗性
resonance	共振
restrained deformation	约束变形
retaining structure	支挡结构
retaining wall	挡土墙
retention	保持量
reveal of reinforcement	露筋
revetment	护坡
rework	返工
right-angle fillet weld	直角角焊缝
rigid analysis scheme	刚性方案
rigid connection	刚接
rigid-elastic analysis scheme	刚弹性方案
rigid foundation	刚性基础
rigid frame	刚架(刚构)
rigid transverse wall	刚性横墙
rigid waterproof layer	刚性防水层

rigid zone	刚域
rigidity	刚度
rigidity of section	截面刚度
rigidly supported continuous girder	刚性支座连续梁
ring beam	圈梁
rise	矢高
rivet	铆钉
riveted connection	铆钉连接
riveted steel beam	铆接钢梁
riveted steel girder	铆接钢梁
riveted steel structure	铆接钢结构
rock discontinuity structural plane	岩体结构面
roller support	滚轴支座
rolled steel beam	轧制型钢梁
roof board；roof plate；roof slab	屋面板
roof bracing system	屋架支撑系统
roof girder	屋面梁
roof system	屋盖
roof truss	屋架
round wire	光圆钢丝

S

safety	安全性
safety class	安全等级
safety classes of building structures	建筑结构安全等级

sapwood	边材
saw-tooth joint failure	齿缝破坏
sampling inspection	抽样检验
sampling scheme	抽样方案
sand-gravel pile	砂石桩法
scarf joint	斜搭接
seamless steel pipe; seamless steel tube	无缝钢管
second moment of area of transformed section	换算截面惯性矩
second order effect due to displacement	挠曲二阶效应
second order elastic analysis	二阶弹性分析
secondary axis	弱轴
secondary beam	次梁
section	截面
section modulus	截面模量(抵抗矩)
section modulus of transformed section	换算截面模量
section steel	型钢
seismic concept design of building	建筑抗震概念设计
seismic fortification criterion	抗震设防标准
seismic fortification intensity	抗震设防烈度
seismic fortification measure	抗震措施
self weight	自重
semi-automatic welding	半自动焊接

separated steel column	分离式钢柱
serious defect	严重缺陷
serviceability	适用性
serviceability limit state	正常使用极限状态
set of high strength bolt	高强度螺栓连接副
setting time	凝结时间
settlement joint	沉降缝
shake	环裂
shaped steel	型钢
shape factor of wind load	风荷载体型系数
shear capacity	受剪承载能力
shear force	剪力
shear modulus	剪切模量
shear plane	剪面
shearing rigidity of section	截面剪变刚度
shearing stiffness of member	构件抗剪刚度
shear strain	剪应变
shear strength	抗剪强度
shear stress	剪应力
shear wall of light wood frame construction	轻型木结构的剪力墙
sheet pile	板桩
shell	壳
shell foundation	壳体基础
shell structure	壳体结构
shield tunneling method	盾构法隧道

short stiffener	短加劲肋
short term rigidity of member	构件短期刚度
shrinkage	干缩
shrinkage crack	收缩缝
silo	储仓
simply supported beam	简支梁
single footing	独立基础
site	场地
site acceptance	进场验收
site load	施工荷载
skylight truss	天窗架
slab	板
slab-column structure	板柱结构
slag inclusion	夹渣
slenderness ratio	长细比
slope protection	护坡
sloping grain	斜纹
slump	坍落度
snow load	雪荷载
snow reference pressure	基本雪压
soda solution grouting	碱液法
soil-lime compacted column	土与灰土挤密桩地基
space rigid unit	空间刚度单元
space structure	空间结构
space suspended cable	悬索
space truss structure	空间网架结构

spacing of bars	钢筋间距
spacing of rigid transverse wall	刚性横墙间距
spacing of stirrup legs	箍筋肢距
spacing of stirrups	箍筋间距
span	跨度
spatial behavior	空间工作性能
specified concrete	特种混凝土
special engineering structure	特种工程结构
spherical steel bearing	球形钢支座
spiral stirrup	螺旋箍筋
spiral weld	螺旋形焊缝
spread foundation	扩展地基
spread foundation	扩展(扩大)基础
square timber	方木
stability	稳定性
stability calculation	稳定计算
stair	楼梯
standard frost penetration	标准冻深
static action	静态作用
static analysis scheme of building	房屋静力计算方案
statically determinate structure	静定结构
statically indeterminate structure	超静定结构
steel	钢材
steel bar	钢筋
steel column component	钢柱分肢
steel column base	钢柱脚

steel fiber reinforced concrete structure	钢纤维混凝土结构
steel hanger	吊筋
steel mesh reinforced brick masonry member	方格网配筋砖砌体构件
steel pipe	钢管
steel plate	钢板
steel plate element	钢板件
steel strip	钢带
steel structure	钢结构
steel support	钢支座
steel tie	拉结钢筋
steel tie bar for masonry	砌体拉结钢筋
steel tube	钢管
steel tubular structure	钢管结构
steel wire	钢丝
step joints	齿连接
stepped column	阶形柱
stiffener	加劲肋
stiffness	刚度
stiffness of structural member	构件刚度
stiffness of transverse wall	横墙刚度
stirrup	箍筋
stone	石材
stone masonry	石砌体
stone masonry structure	石砌体结构

story with outriggers and/or belt member	加强层
straight-line joint failure	通缝破坏
straightness of structural member	构件垂直度
strain	应变
strand	钢绞线
strength	强度
strength classes of masonry unit	块体强度等级
strength classes of mortar	砂浆强度等级
strength classes of structural steel	钢材强度等级
strength classes of structural timber	木材强度等级
strength classes(grades)of concrete	混凝土强度等级
strength classes (grades) of steel bar	普通钢筋强度等级
strength of structural timber parallel to grain	木材顺纹强度
stress	应力
strip foundation	条形基础
strong axis	强轴
structural composite lumber(SCL)	结构复合木材
structural concrete column	混凝土构造柱
structural glued-laminated timber	胶合木结构
structural plywood	结构胶合板
structural system composed of bar	杆系结构
structural system composed of plate	板系结构

structural wall	结构墙
structural wood-based panel	木基结构板材
structure	结构
stud	墙骨
stud welding	焊钉(栓钉)连接
superposed beam	叠合梁
surface course	面层
suspended crossed cable net	双向正交索网结构
suspended structure	悬挂结构
swan and round timber structures	方木和原木结构
swan and round timber structures	普通木结构

T

tall building	高层建筑
tangential strain	剪应变
tangential stress	剪应力
temperature action	温度作用
tensile capacity	受拉承载能力
terrain roughness	地面粗糙度
test assembling	预拼装
the smallest assembled rigid unit	小拼单元
thickness of concrete cover	混凝土保护层厚度
thickness of section	截面厚度

thin shell	薄壳
three hinged arch	三铰拱
tie bar	拉结钢筋
tied framework	绑扎骨架
timber roof truss	木屋架
timber structure	木结构
torque	扭矩
torsional capacity	受扭承载能力
torsion rigidity of section	截面扭转刚度
torsion stiffness of member	构件抗扭刚度
transfer member	转换结构构件
transfer story	转换层
transient situation	短暂状况
transverse horizontal bracing	横向水平支撑
transverse stiffener	横向加劲肋
transverse weld	横向焊缝
transversely distributed steel bar	横向分布钢筋
trapezoid roof truss	梯形屋架
triangular pyramid space grid	三角锥体网架
triangular roof truss	三角形屋架
tributary area	从属面积
trussed arch	椽架
trussed rafter	桁架拱
total breadth of structure	结构总宽度
total height of structure	结构总高度
total length of structure	结构总长度

toweling course	找平层
truss	桁架
truss plate	齿板
tube in tube structure	筒中筒结构
tube structure	筒体结构
twist	扭弯
two hinged arch	双铰拱
two sides(edges)supported plate	两边支承板
type inspection	型式检验
type of roof or floor structure	屋盖、楼盖区别

U

ultimate compressive strain of concrete	混凝土极限压应变
ultimate deformation	极限变形
ultimate limit state	承载能力极限状态
ultimate strain	极限应变
undercut	咬边
under layer	垫层
underground waterproof engineering	地下防水工程
uniform cross-section beam	等截面梁
unit weight	重力密度(重度)

unseasoned timber	湿材
upper flexible and lower rigid complex multistory building	上柔下刚多层房屋
upper rigid lower flexible complex multistory building	上刚下柔多层房屋

V

value of decompression prestress	预应力筋消压预应力值
value of effective prestress	预应筋有效预应力值
variable action	可变作用
variable load	可变荷载
verification of serviceability limit state	正常使用极限状态验证
verification of ultimate limit state	承载能极限状态验证
vertical bracing	竖向支撑
vibration	振动
visual examination of structural member	构件外观检查
visual examination of structural steel member	钢构件外观检查
visual examination of weld	焊缝外观检查
visually stress-graded lumber	木材木测分级

W

wall	墙
wall beam	墙梁
wall frame	壁式框架
warping	翘曲
warping rigidity of section	截面翘曲刚度
water tower	水塔
water/cement ratio	水灰比
weak region of earthquake-resistant building	抗震建筑薄弱部位
web plate	腹板
weld	焊缝
weld crack	焊接裂纹
weld defect	焊接缺陷
weld roof	焊根
weld toe	焊趾
weld ability of steel bar	钢筋可焊性
welded framework	焊接骨架
welded steel beam; welded steel girder	焊接钢梁
welded steel pipe	焊接钢管
welded steel structure	焊接钢结构

welding connection 焊缝连接
wind fluttering factor 风振系数
wind load 风荷载
wind reference pressure 基本风压
wind-resistant column 抗风柱
wall-slab structure 墙板结构
wind vibration 风振
wood preservative 木材防护剂
wood-based panel 木基复合板材
wood-based structural-use panel 木基结构板材
wood-frame construction 轻型木结构
wood roof decking 屋面木基层

yield strength 屈服强度

Reference

[1] 李嘉. 专业英语[M]. 北京:人民交通出版社, 2012.

[2] DAVID M W. Civil engineering, a very short introduction[M]. Oxford:Oxford University Press, 2012.

[3] PETER A C. Civil engineering materials [M]. Oxford: Butterworth-Heinemann Publications, 2016.

[4] SAMUEL L. Introduciton to civil engineering system [M]. Hoboken:John Wiley &Sons. Inc. , 2014.

[5] GIOVANNI C M, LEN H. Introduction to construction project engineering[M]. New York: House Publishing, 2018.

[6] 魏华, 梁旭黎. 土木工程专业英语[M]. 西安:西安交通大学出版社, 2015.

[7] 李丰. 土木工程专业英语[M]. 北京:北京理工大学出版社, 2016.

[8] 王静峰. 土木工程专业英语[M]. 北京:机械工业出版社, 2019.

[9] 郭仁东, 孙雨明, 荆辉. 土木工程专业英语[M]. 北京:人民交通出版社, 2015.

[10] 姜晨光. 土木工程专业英语教程[M]. 北京:化学工业出版

社, 2013.
[11] 王佐才, 王顶堂. 土木工程专业英语[M]. 武汉:武汉大学出版社, 2013.
[12] 戴俊, 刘纯中. 土木工程专业英语[M]. 北京:机械工业出版社, 2008.
[13] 王清标,李庆学. 土木工程专业英语[M]. 北京:机械工业出版社, 2012.
[14] 苏小卒. 土木工程专业英语[M]. 上海:同济大学出版社, 2011.
[15] 吴轶. 土木工程专业英语[M]. 武汉:武汉大学出版社, 2015.
[16] 贾艳敏. 土木工程专业英语[M]. 哈尔滨:哈尔滨工业大学出版社, 2014.
[17] 段兵廷. 土木工程专业英语 [M]. 武汉:武汉工业大学出版社, 2013.
[18] 张子新, 胡欣雨. Underground structure[M]. 北京:中国建筑工业出版社, 2009.
[19] 王俊, 齐玉军. Introduction to civil engineering[M]. 北京:中国建筑工业出版社, 2021.
[20] 黄莺. Introduction to civil engineering[M]. 北京:中国建筑工业出版社, 2020.
[21] 刘娟, 李春香. 土木工程实用英语[M]. 成都:西南交通大学出版社, 2015.
[22] CHEN W F,DUAN L. Bridge engineering handbook[M]. New York:CRC Press, 2020.
[23] KAROLY Z. Structural analysis of multi-storey buildings[M]. New York:CRC Press, 2020.
[24] XIAO M. Geotechnical engineering design[M]. Oxford:Library of Congress Cataloging in Publication, 2015.

[25] BARBARA G, ROBERT A. How to write and publish a scientific paper [M]. California: Greenwood Publications, 2016.

[26] 王圣林, 刘运生. 英汉土木建筑工程词汇手册[M]. 北京: 中国建筑工业出版社, 2005.